Easy Container Gardening in Any Climate

The Ultimate Guide to Grow Your Own
Organic Herbs and Vegetables, Create a Sustainable Environment,
Maximize Yields, and Utilize Any Size Space

Kiley Manning

ISBN: 978-1-923422-09-4

Table of Contents

Introduction

I've always wondered what it would be like to have my own garden—a little oasis for myself. Every time I watched videos online of people setting up their own paradise, I have to admit I started desiring to add a bit more greenery and color to my concrete surroundings.

Even after spicing up my apartment with colorful decorations and trinkets, I later found myself wanting to do something new and nurturing something that's a bit more natural.

So, I began my container gardening journey on my little balcony in a bustling city, and what started as curiosity and the want for a hobby slowly blossomed into a passion that transformed both my living space and lifestyle. Tasting a sun-warmed tomato, freshly picked from my balcony, was a simple yet delightful experience. This personal journey highlighted the profound benefits of growing my own food and fostered a deeper connection with nature.

But even you can transform a small living space into a thriving garden with tomatoes, basil, and lettuce—so I wrote a book about everything I've learned along the way!

Container gardening maximizes limited space, whether you're living in a high-rise apartment or a suburban home, creating a refreshing green and natural area in the confines of an urban environment. It's a practical and sustainable way to grow produce and beautiful and fragrant flowers for your living space.

As you dive into this book, you'll learn the essentials of choosing the right containers, understanding soil composition, and mastering the art of proper watering. You'll progress to navigating various climate conditions and discovering innovative gardening technologies.

This book also goes beyond planting basics, diving into sustainable practices that help you create an eco-friendly garden. Each chapter is brimming with practical tips, step-by-step guides, and DIY projects that are both enjoyable and functional. You will find that this book simplifies even the most complex gardening techniques.

This is a hobby that's more than just growing plants and self-sustainability; it's about cultivating well-being. As urban areas expand and living spaces shrink, turning to our balconies, windowsills, and rooftops can enhance our homes, improve our mental and physical health, and contribute to urban sustainability.

Just think about the potential of that sunny spot in your living room, the bare corner of your patio, or the unused space on your kitchen counter. With this book, you can transform these areas into a thriving garden.

Today is the perfect day to start your adventure in container gardening, and I am here to guide you every step of the way.

- *Kiley Manning*

Like people, plants respond
to extra attention

- H. Peter Loewer

Chapter 1

Container Gardening Basics

The first step in starting your container gardening journey is learning to pick and choose the right pots and plants, understanding the use of different soil mixes, and planning when and where to set your gardening space.

1.1 Choosing the Right Containers

Using the right containers means being able to grow healthier plants while enhancing the beauty of your space—this would mean keeping the material, volume, shape, drainage, and even the style or design in mind.

Container Material

Picking the right material not only influences the aesthetics of your garden but also the health and growth of your plants. Each material comes with its unique set of advantages and drawbacks, impacting factors like moisture retention, durability, and suitability for different environments.

Having a better grasp of the unique attributes of these materials can give you the ability to craft an informed decision that is finely tuned to your gardening preferences.

Commonly used materials for pots include plastic, clay, metal, wood, and fabric:

- **Plastic Pots**
  - **Advantages**: Lightweight, affordable, retains moisture well, and works best in hotter climates.
  - **Disadvantages**: Easier to shift and topple down. Easily becomes brittle—cracks in colder climates and degrades faster with constant sun exposure.
- **Clay Pots**
  - **Advantages**: Porous, allowing better air circulation to the roots, and natural water drainage helps prevent diseases like root rot.
  - **Disadvantages**: Heavier, more fragile, more likely to break if dropped.
- **Metal Containers**
  - **Advantages**: Durable, offers a sleek look. Lasts longer than plastic pots.
  - **Disadvantages**: Conducts heat that could potentially overheat the soil on hot days and may require internal insulation or strategic placement to avoid direct sunlight during peak hours.
- **Wooden Containers**
  - **Advantages**: Has a natural look that blends well with outdoor environments and stabilizes temperature better for plant roots.
  - **Disadvantages**: Can rot over time if not properly treated and may need regular maintenance and sealing.
- **Fabric Pots**
  - **Advantages**: Lightweight, flexible material, excellent drainage, and air circulation; promotes

root pruning and is easier to store when not in use.

- **Disadvantages**: Less durable, may dry out more quickly, requiring frequent watering.

Container Volume and Root Space

A common mistake for beginners is choosing pots that are either too small or too big for their plants, leading to overcrowded roots and stunted growth. Having too little space for your plant to grow into can cause the roots to become cramped and tangled, while too much space causes roots to spread too far—both issues result in poor nutrient intake for your plants.

There are three container sizes you can choose from: small, medium, and large. For every plant, there are suitable pot sizes that can help them thrive better depending on their root system and needs.

Small containers, typically up to 1-3 gallons (3.87-11.35 L), are best suited for shallow-rooted plants such as herbs and small annuals.

These plants have modestly sized root systems that can flourish even within limited soil volumes. However, due to the restricted space, they require frequent watering and fertilizing to ensure they receive adequate nutrients.

Plants for small containers:

- **Herbs**: Basil, parsley, thyme, and chives can grow well in small containers. For instance, a 1-gallon (3.87 L) pot is sufficient for a single basil plant.
- **Annual Flowers**: Marigolds and petunias are examples of small annuals that can thrive in small containers, adding color to your garden without requiring extensive root space.

Medium containers are usually between 3-7 gallons (11.35-26.49 L) and are ideal for compact vegetables and flowering plants.

These containers provide a balance between manageable size and adequate root space, making them versatile for most plants.

Plants for medium containers:

- **Compact Vegetables:** Bush beans, dwarf tomatoes, and peppers can thrive in medium containers. A 5-gallon (18.92 L) container can comfortably support a pepper plant.
- **Flowering Plants:** Geraniums, zinnias, and impatiens grow well in medium containers, offering vibrant blooms and healthy growth.

Large containers, which are 10 gallons (38.75 L) or more, are necessary for larger plants and shrubs. These containers offer ample room for root expansion, ensuring that the plants have enough space to access nutrients and water. They're also less prone to rapid drying out, providing a more stable growing environment.

Plants for large containers:

- **Large Vegetables:** Tomato plants, especially indeterminate varieties, and eggplants require large containers. A 10-15-gallon (37.85-56.78 L) pot is the perfect size for a tomato plant to flourish in abundant cultivation.
- **Shrubs and Small Trees:** Dwarf citrus trees, blueberries, and ornamental shrubs thrive in large containers, allowing for significant root growth and healthy development.

Container Shape

Other than size, the shape of the container can influence root growth patterns and the overall stability of the plant. Different shapes are suitable for different types of plants.

Tall, narrow containers are suitable for deep-rooted plants, allowing their roots to grow downward efficiently.

Plants for tall, narrow containers:

- **Deep-rooted Vegetables:** Carrots, parsnips, and beets benefit from tall, narrow containers that accommodate their long roots.
- **Ornamental Plants:** Plants like bamboo and certain ornamental grasses thrive in tall containers that mimic their natural growing conditions.

However, these containers can be prone to tipping over, especially in windy conditions, due to their height and narrow base. Wide, shallow containers provide extra stability; these are ideal for plants with shallow root systems.

These containers offer a broad surface area, making it easier for plants to access sunlight and for gardeners to arrange multiple plants in a single container.

Plants for wide, shallow containers:

- **Herbs:** Oregano, thyme, and mint grow well in wide, shallow containers, spreading their roots horizontally.
- **Succulents:** Varieties like hens and chicks, sedum, and echeveria are well-suited to wide, shallow containers that provide excellent drainage and air circulation.

Container Layering and Transplanting

Layering your containers is a good way to organize your container garden space—especially in keeping track of your different-sized plants and pots.

For deep containers, consider layering different types of soil. Start with coarse materials like pebbles at the bottom for drainage, followed by a layer of rich compost, and top it with potting mix.

As for transplanting seedlings to other containers, handle them gently by the leaves, not the stems, to avoid damage. Acclimate the seedlings by introducing them to outdoor conditions bit by bit before transplanting.

Creatively Repurposed Containers

Upcycling old glass jars, plastic Tupperware, and metal tins can add character and charm to your container garden space. However, check if these repurposed containers are safe for plants by ensuring that these unique pots have not been treated with potentially toxic or harmful chemicals.

To prepare these materials for any potential contaminants, you must first clean them properly by removing small organic and/or chemical debris (e.g., food, cream, soap, etc.).

Clean off any foreign material you find in the potential plant container; dishwashing is fine as long as you place your plastics in the top rack and glass and metals in the bottom rack.

If you're going for the more manual approach, you can scrub off the scraps and bits and soak them in warm, soapy water for about 5-10 minutes, which is enough to get rid of food and lotion oils stuck in them.

When removing labels, soaking your container in warm water for about five or six minutes usually softens the label's adhesive. For the more irritating labels that just won't come off, you can try putting together a homemade "*goo remover*" by mixing baking soda, coconut oil, and essential oils like sweet orange or lemon.

Next, once you're done with the first phase of cleaning your containers, you can proceed with the next step: sterilizing your containers to clear them of any plant-toxic substances.

For glass and metal containers, put them in a large pot filled with water and bring to a boil for about 15 minutes. Remove with sterilized tongs afterward and fully air dry them before use.

Plastics, on the other hand, will need to be carefully soaked in 3% hydrogen peroxide for 10–15 minutes and quickly rinsed off. Leave to dry in a clean area. Alternatively, you can spray them with 70% isopropyl alcohol and leave them to dry.

Some of these containers can contain unpleasant smells and might stay that way even after cleaning and sterilizing. When this happens, you can leave these containers out in the sun to get rid of any odors.

You can also sprinkle a bit of baking soda in the now clean and dry containers, put their lids on, leave them for a full day (24 hours), and finally rinse and dry them.

Proper Drainage

Having effective drainage is important to prevent water from forming pools at the base of the pot, which can cause root diseases like root rot.

The first step in making sure that your drainage is adequate is to check that your containers have adequate drainage holes. If your container doesn't have holes, you can carefully drill holes in the bottom of the container. For ceramic or clay pots, use a masonry drill bit, and for plastic, a standard bit will do.

Double-potting would be ideal for indoor container plants and can be easily done by having a decorative container as a cover for a smaller pot that has drainage holes. This method also makes it easier to move plants around.

Now that your containers have drainage holes, you can start layering the bottom of your containers with gravel, pebbles, or broken ceramics to improve water flow by preventing soil from blocking the drainage holes.

Here's how to do it:

- **Gravel or Pebbles:** Place a 1-2-inch (2.54-5.08 cm) layer at the bottom. This layer allows excess water to drain quickly and prevents soil from compacting around the drainage holes.
- **Broken Ceramics or Pottery:** Recycle broken pots or tiles by using them to create a drainage layer. They work similarly to gravel and pebbles.

Container Style Considerations for Plant Needs

This could be the first—or last—thing to consider when you're picking the right containers in preparation for your personal container gardening.

When selecting stylish and functional containers, things like color and mobility can actually affect your plants when you're balancing aesthetics with the specific needs of your plants.

Other container considerations:

- **Color**: Choosing the right colors can have some impact on a plant's growth.
    - Light-colored containers reflect heat and can keep the soil cooler, which is beneficial for plants that prefer cooler root zones.

    - Dark-colored containers absorb heat, which can be advantageous for heat-loving plants but may require more frequent watering.
- **Mobility:** Consider the mobility and weight of your containers.
    - Heavy ceramic or concrete pots are stable but difficult to move. Adding wheels will make it easier
    - Adding wheels or using lighter materials can make it easier to rearrange your garden as needed.
- **Design Consistency:** Finally, pick containers that match the overall theme of your garden.
    - For a rustic look, opt for wooden or terracotta pots.
    - For a modern aesthetic, sleek metal or monochromatic plastic pots can be more fitting.

The next page contains a short and quick container selection checklist that you can use for future reference.

Container Selection Checklist

Make sure you're using the right material that suits your climate and garden's aesthetic.

- **Hotter Climates:** Clay and fabric pots
- **Cooler Climates:** Metal pots
- **All-Around:** Plastic and wooden pots
    - Note: Plastic is more fragile; cracks in colder climates

Size your pots according to your plant's mature root system

- Small
- Short
- Narrow
- Medium
- Tall
- Wide
- Large

Look for drainage holes in your pots or drill new holes.

Check if the container is plant-safe, clean, and non-toxic.

Plan out the overall style you want to achieve for your container garden by paying attention to these elements:

- Color and Material
- Mobility
- Design

1.2 The Perfect Soil Mix

After choosing the right containers for your garden, the next step is to select the right soil for your plants to put into. If the container is a plant's home, then the soil serves as the foundation, providing both physical support and the vital nutrients necessary for healthy plant growth.

A well-formulated soil mix for container gardening must balance excellent drainage and adequate water retention, and in doing so, this ensures that plant roots get sufficient moisture and nutrients while preventing waterlogging, which can lead to different diseases—like root rot.

Pre-made vs. Homemade Soil Mixes

Most gardeners will have to choose between preparing their own soil mix or using pre-made mixes available at garden centers. Each option has its benefits, and understanding these can help you make the best decision for your plants.

Pre-made mixes are convenient and are often formulated with the right balance of ingredients for good drainage and moisture retention. When using pre-made mixes, consider adding organic fertilizers or specific amendments tailored to the needs of your plants.

A standard pre-made potting mix might include:

- **Peat Moss:** Provides moisture retention and aeration.
- **Perlite:** Enhances drainage and prevents soil compaction.
- **Vermiculite:** Holds water and nutrients, releasing them slowly to plant roots.

However, this pre-made mix might lack specific nutrients crucial for certain plants. To boost its nutritional content, you can add:

- **Organic Fertilizers:** Gathered from natural sources like fish emulsion or natural techniques like compost tea.
- **Slow-Release Fertilizers:** A type of organic fertilizer, like Osmocote, which provides a steady supply of nutrients over time.

Homemade mixes allow for customization and are perfect for general gardening tasks. You can adjust components based on your plant's needs.

Below are some simple recipes for nutrient-rich, well-draining soil foundations for a wide range of plants:

- **Two parts compost:** Rich in nutrients and beneficial microorganisms. Boosts nutrient content for heavy feeders.
- **One part perlite:** Enhances drainage and aeration, especially for succulents.
- **One part peat moss or coconut coir:** Provides moisture retention and improves soil structure.

Advanced Soil Amendments

For more specialized plants or improved soil health, consider these advanced amendments:

- **Coconut Coir:** An ecologically sound substitute to peat moss, this improves water retention and aeration. Ideal for tropical plants that require consistent moisture.
- **Biochar:** Enhances soil fertility and water retention while supporting beneficial microbial activity. Suitable for vegetable gardens to boost overall plant health and yield.

- **Worm Castings:** A rich source of nutrients that improves soil structure and boosts plant growth. Enriches the soil for fruiting plants like tomatoes and peppers.
- **Mycorrhizae:** Beneficial fungi that increase plants' ability to absorb water and nutrients. Particularly beneficial for perennials and trees.
- **Green Manure:** Young plants tilled back into the soil add organic matter and nutrients. Legumes like clover or vetch improve soil nitrogen levels.

Custom Soil Blends

These soil blends are precisely formulated for the unique needs of specific plants and can greatly aid their growth and health.

Each basic garden soil component contributes differently to soil texture, drainage, and nutrient-holding capacity:

- **Sandy Soil:** Has great drainage but may need more organic matter to hold nutrients.
- **Silt Soil:** Smooth and slippery when wet. Holds moisture and nutrients well but can compact easily.
- **Clay Soil:** Holds nutrients well but may require soil amendments to improve drainage.

Once you're familiar with the basic garden soil components, you can start combining and preparing specialized soil blends with additional amendments.

Providing the right soil blend for your plants can be detrimental to their overall health, especially when it comes to fulfilling your plants' more specific maintenance needs.

Here are some plants that would require some special attention:

- **Cacti and Succulents:** Thrive in quick-draining soils with a higher proportion of sand and perlite.
- **Moisture-Loving Plants:** Ferns benefit from a mix with more coir or peat moss.
- **Acid-Loving plants:** Blueberries and azaleas would prefer acidic soil with a lower pH, which can be achieved by amending soil with pine needles or sulfur.

Sustainability in Soil Management

Some gardeners use peat moss to improve their soil, especially when the soil is not acidic enough for certain plants. However, harvesting peat moss from peat bogs can release stored carbon or greenhouse gases into the atmosphere. This contributes to climate change. Minimizing the use of peat moss is a step towards sustainable gardening practices.

These days, it's better to choose environment-safe alternatives like coir or composted bark, as they provide similar benefits that preserve the ecosystem.

Choosing renewable soil amendments, such as locally produced compost or algae-based fertilizers, supports local economies and reduces overall environmental impact. Here are some notable techniques used in managing and maintaining soil quality:

1. **Aeration:** Regularly aerate the soil to maintain good airflow and prevent compaction. Use tools like a hand fork or even a normal fork you don't use anymore to loosen the soil gently.

2. **Moisture Retention:** Incorporate water-retentive granules or gels, especially for container plants that require consistent moisture. These additives help maintain a stable moisture level while reducing the frequency of watering.
3. **Nutrient Cycling:** Rotate crops or vary plant types within containers to prevent nutrient depletion. Different plants have varying nutrient needs, and cycling helps maintain soil fertility.
4. **Soil Testing:** Regular soil testing helps understand the current condition and guides necessary adjustments. A majority of plants thrive within the pH range of 6.0 and 7.0. If your soil tests are acidic, adding garden lime can raise the pH; if it's too alkaline, sulfur or aluminum sulfate can lower it.
5. **Balanced Nutrient Levels:** Symptoms like yellowing leaves or stunted growth often indicate nutrient deficiencies that can be corrected with targeted amendments.

By taking the time to understand and manipulate the soil in your containers, you set the stage for a lush and productive container garden.

Tailoring soil blends and amendments to suit specific plant needs and environmental conditions transforms gardening from a pastime into a fine-tuned craft. This method guarantees that your plants bask in optimal growing conditions, fostering robust growth, resplendent blooms, and abundant yields.

1.3 Climate and Weather Adaptations

Depending on where you live, understanding your local climate and weather is fundamental to the success of your container garden.

Different plants thrive under specific climates, and aligning your plant choices with your local environment ensures a thriving garden. This next section combines insights on climate zones, microclimates, and seasonal weather adaptations to provide a comprehensive guide for container gardening.

Climate Zones

Familiarize yourself with climate zones categorized by temperature, humidity, and seasonal changes. These zones help gardeners select plants most likely to succeed in their region. Plants like succulents thrive in the arid climates of the Southwest, and ferns flourish in the moist, shaded environments of the Pacific Northwest.

Checking the USDA Plant Hardiness Zone Map can be a great starting point. This map divides North America into 13 zones that were established based on the average annual minimum winter temperature.

Knowing your zone helps you select plants that can survive and thrive in your climate. Consider the microclimate of your specific gardening space. Microclimates are small areas where the climate differs from the surrounding area due to factors like proximity to buildings, bodies of water, or landscape features.

For instance, a garden next to a south-facing wall may be more exposed to higher temperatures, making it suitable for plants that need more warmth and sunlight. Factors that influence microclimates include:

- **Proximity to Structures**: Buildings can provide shelter from wind and reflect heat, creating warmer spots.
- **Bodies of Water**: Lakes, ponds, or even large water features can moderate temperatures, keeping the area cooler in summer and warmer in winter.
- **Elevation Changes**: Even small changes in elevation can affect temperature and wind exposure.

Sunlight, Shade, and Wind Exposure

Understanding the sunlight patterns in your garden space is crucial for selecting suitable plants. Nearly all plants require a specific amount of sunlight each day:

- **Full Sun**: At least 6-8 hours of direct sunlight. Ideal for vegetables like tomatoes and peppers.
- **Partial Sun/Shade**: 3-6 hours of direct sunlight or filtered light. Suitable for herbs like parsley and chives.
- **Full Shade**: 3 or fewer hours of direct sunlight. Perfect for shade-loving plants like hostas and ferns.

Use a sunlight calculator or observe your garden throughout the day to determine how much light different areas receive. Adjust your plant choices based on these observations to ensure they get the appropriate amount of light.

Other than sunlight, you have to consider wind exposure along with sun exposure. Having your plants exposed to a cool breeze and a bit of sun can be beneficial to the overall health of your container garden. But if you live in a windy and hot area. In urban environments or on high balconies, plants can suffer from dehydration or physical damage.

To reduce these effects:

- Place your containers where buildings or structures can act as windbreaks.
- Choose wind-resistant plants like grasses, native species, and hardy shrubs.
- Use heavier containers to prevent your garden from being toppled by strong gusts.

Additionally, reflective surfaces like windows or walls can increase light and heat, benefiting sun-loving plants. On the other hand, areas with less exposure might be cooler and more humid, which can support shade-loving plant and vegetable species.

Temperature, Humidity, and Weather

In container gardening, container plants are more susceptible to temperature fluctuations, especially if their roots are exposed. To protect your plants:

- Insulate pots with bubble wrap or fabric to shield roots from extreme temperatures.

- Use light-colored containers to reflect heat and cool the soil in hot weather.
- Move containers to sheltered locations during extreme weather conditions, such as heat waves or frosts.

Humidity levels can also affect plant health. Some plants, like tropical varieties, thrive in high humidity, while others, such as succulents, prefer dry conditions. Misting plants or using humidity trays can help maintain the right moisture levels for humidity-loving plants. Integrating knowledge of climate zones, sunlight, shade, wind, and exposure into your gardening ensures a thriving, harmonious ecosystem. As you nurture your container garden, becoming attuned to the nuances of weather and its impact on plant health is essential.

Each season carries its own set of challenges, from the scorching heat of summer to the biting cold of winter, so adapting your gardening practices accordingly can significantly develop resilience and boost the vitality of your plants.

Adapting to Hot and Cold Climates

During periods of intense heat, container plants can suffer from overheated soil, which can damage roots and hinder growth. To mitigate this:

- **Positioning**: Place containers where they receive morning sunlight but are shaded during the hottest part of the day. For example, noontime, when the sun is at its peak, would easily harm your shade-loving plants.
- **Shade Cloths**: Use shade cloths to provide relief from intense sun exposure by hanging them up like a canopy above your plants.

- **Mulching**: Generously add a thick layer of organic mulch, such as straw or bark, to the surface of the soil to insulate and retain moisture.
- **Watering Techniques**: Water your plants either at dawn or later in the evening to conserve moisture and nourish the roots effectively.

If you live in colder climate areas, during cold snaps, protect your plants by:

- **South-Facing Walls**: Move containers to a south-facing wall to benefit from the retained heat of the building.
- **Frost Covers**: Use frost covers overnight to shield plants from sudden temperature drops.
- **Insulation**: Wrap containers with insulating materials such as bubble wrap or burlap.
- **Windbreaks**: Set up barriers like fences or screens to reduce wind exposure and minimize cold damage.

Seasons and Unexpected Weather Events

Other than accounting for weather and climate in your area, the changing seasons will also have to be given just as much attention. Adjusting by the season should be a part of your garden's maintenance routine:

- **Watering**: As temperatures drop, reduce watering frequency to match the decreased evaporation rate.
- **Plant Feeding**: Adjust your feeding schedule by reducing high-nitrogen fertilizers and shifting towards phosphorus-rich feeds.
- **Lifecycle Management**: Understand the lifecycle and growth patterns of your plants. Introduce or retire plants by their optimal growing conditions for each season.

Not all plants are year-round growers, so here are some examples of plants that you can try growing for these four seasons:

- **Spring**: Pansies, violas, lettuce, peas.
- **Summer**: Tomatoes, basil, marigolds, cucumbers.
- **Autumn**: Mums, kale, ornamental grasses, asters.
- **Winter**: Evergreens, hellebores, winter pansies, holly.

Some areas in the US, especially southern areas like Florida, Louisiana, and Texas, are susceptible to unexpected weather events. If you live in an area where the weather can take a drastic turn, creating an emergency plan for your garden will involve the following:

- **Risk Identification**: Identify potential risks such as high winds or heavy rains.
- **Mitigation Strategies**: Secure loose containers with weights or ties.
- **Supplies**: Keep extra mulch to protect against sudden cold and sandbags to fortify against flooding.
- **Emergency Shelters:** Set up temporary shelters, such as pop-up greenhouses or portable cold frames.

Additional Weather and Climate Adaptation Tips:

- **Monitoring Weather Forecasts**: Stay informed about upcoming weather conditions.
- **Plant Selection**: Choose varieties known for their hardiness and adaptability to local weather conditions.
- **Soil Preparation**: Enhance soil drainage by incorporating organic matter and ensuring proper container drainage to prevent waterlogging during heavy rains.

1.4 Planting Techniques and Timing

Creating a thriving container garden involves seasonal timing, proper watering, and mulching. Each element is crucial for the health of your plants, affecting root development and their ability to absorb water and nutrients effectively.

Seasonal Timing

Timing your planting is crucial for garden success. Most vegetables, herbs, and flowers have specific growing seasons. Understanding your local climate helps you schedule and adjust your planting activities for optimal growth conditions.

- Cool-season vegetables like lettuce and spinach should be planted in early spring or fall.
- Warm-season crops like peppers and cucumbers thrive after the last frost.
- Flowers such as pansies and primroses are ideal for early spring.
- Herbs vary; basil prefers warm weather, while cilantro does better in cooler temperatures.

Below are some of the more important timing techniques and dates that you need to take note of:

- **Succession Planting:** To maximize yield, practice succession planting. This involves planting a new crop immediately after harvesting the previous one, ensuring continuous production.
- **Staggered Planting:** Plant in stages to spread out the harvest period. For example, plant a batch of lettuce seeds every two weeks for a steady supply.
- **Frost Dates:** Know your area's first and last frost dates. Use this information to strategically start seeds indoors and subsequently transplant them outside.

Setting Up Your Space

Creating a flourishing container garden starts with assessing and utilizing your available space. Whether you have a balcony, a patio, or indoor areas, each setting offers unique opportunities and challenges. Balconies present challenges like limited space and wind exposure, but these areas also offer great opportunities for elevated gardens.

Consider the weight of your containers, especially when wet, to avoid possible structural issues—lightweight materials like fiberglass or resin are ideal for balconies. Secure your containers to prevent accidents from strong winds by adding railing planters that can be attached securely and efficiently use space while enhancing privacy.

Tall grasses or bamboo can also act as natural privacy screens.

Important details for balconies:

- **Vertical Gardens:** Utilize vertical space with wall-mounted planters or hanging baskets to maximize planting areas.
- **Microclimates:** Pay attention to microclimates on your balcony. For instance, corners might be more sheltered and retain heat, while exposed areas might be windier and cooler.

Outdoor spaces like Patios and small yards allow for more ground contact, which is ideal for larger plants. You can maximize these spaces with vertical gardening using climbing frames, trellises, or wall-mounted containers. Raised beds or tiered planters can also save space and ease the strain on your back and knees.

Plan your layout to ensure all plants receive adequate light by considering the sun's movement and arranging taller plants so they don't cast shadows on shorter, sun-loving plants.

Important details for outdoors:

- **Mobility:** Use wheeled plant caddies or movable containers to shift plants according to seasonal light changes.
- **Mixed Plantings:** Combine ornamental and edible plants for a multifunctional garden space that's both beautiful and productive.

For indoor container gardening, you will need to understand your home's conditions, such as light levels, humidity, and temperature. Use complementary lighting like grow lights in low-light areas, consider using humidifiers for tropical plants, and provide dry conditions for succulents.

If you want to improve indoor air quality with plants that thrive in lower light and drier air conditions, choose adaptable plants like pothos, snake plants, and peace lilies.

Important details for indoors:

- **Natural Light Maximization:** Place plants near windows that receive the most sunlight. South-facing windows typically provide the most light.
- **Group Plants:** Grouping plants together can help increase humidity levels around them, creating a more favorable microclimate.

By focusing on the specific requirements and solutions for each type of space, you can create a vibrant and productive garden.

Watering Techniques

Watering your plants in your container garden differs slightly from watering garden beds. Container gardens dry out faster than other gardens, which means they require more frequent watering than usual. The goal is to moisten the entire root zone.

First, check if watering is needed by feeling the soil about an inch below the surface; if it's dry, it's time to water. Next, water until liquid flows out of the drainage holes to ensure the lower roots are reached.

However, avoid overhead watering in the evenings to prevent wet foliage overnight, which can lead to fungal diseases. Watering in the morning helps plants withstand the daytime heat.

If you're having trouble finding time to water your plants, setting up self-watering containers can be an excellent solution for consistent moisture.

- **Water Retention:** Use water-retaining crystals or hydrogels mixed into the soil to help retain moisture.
- **Mulching:** Reduce evaporation and suppress weeds by adding a layer of mulch on top of the soil.
- **Drip Irrigation:** Consider setting up a drip irrigation system to consistently and slowly water the roots directly, reducing water waste and preventing overwatering.
- **Rainwater Harvesting:** Use collected rainwater for irrigation. It's often better for plants as it doesn't contain the chemicals found in tap water.

Mulching and its Benefits

Mulching is vital in container gardening for maintaining soil moisture and regulating temperature. A good mulch can reduce watering needs by up to 50% and control weed growth while regulating temperature of the soil during the hotter and cooler months, creating a stable and well-insulated environment for roots. Organic mulches like straw, shredded bark, or coco coir add valuable organic matter to the soil as they decompose.

To add mulch, simply apply a 2-inch (5.08 cm) layer of mulch on top of the soil, avoiding contact with plant stems to prevent rot.

Benefits of mulching

- **Soil Erosion Prevention:** Mulching helps prevent soil erosion by reducing the effect of rain on the surface of the soil. This is particularly beneficial in container gardening, where soil can be easily displaced.
- **Enhanced Soil Structure:** Organic mulches decompose, improving the structure of the soil by adding to the organic matter content, which enhances water retention and root penetration.
- **Pest Control:** Certain mulches, like cedar and pine bark, can help deter pests. They release natural oils and compounds that are unpleasant for many insects.
- **Aesthetic Appeal:** Mulch can also enhance the visual appeal of your container garden. Using decorative mulches like colored wood chips or gravel can add a finished look to your garden setup.

1.5 Essential Tools and Gear

Having the right tools and protective gear can make all the difference in your container gardening experience. This section will provide you with a list of essential items you'll need for efficient and enjoyable gardening, explaining what each item does and why it's useful.

Basic Tools

You'll need a set of basic tools to prepare yourself for container gardening. Here are some tools that can help you with planting, watering, and maintaining your plants:

- **Large-Spouted Watering Can:** A container with a handle and a large spout for watering plants. The large spout ensures a gentle and even water flow, preventing soil disruption and ensuring water reaches the plant roots effectively. A sturdy and long-lasting watering can would be made of galvanized steel or durable plastic.
- **Scissors**: A pair of sharp, small cutting tools. They are ideal for delicate tasks such as trimming small plants, deadheading flowers, and harvesting herbs without damaging the plants. Choose scissors with ergonomic handles and stainless-steel blades for durability and sharpness.
- **Pruners**: Also known as secateurs, these are strong scissors for cutting plant stems and branches. They are essential for pruning and cutting back larger stems and branches, keeping plants healthy and well-shaped. Ergonomic handles and stainless-steel blades would be ideal. Use regularly and cut at a 45-degree angle.
- **Trowel**: A small hand tool with a pointed, scoop-shaped metal blade and a handle that could be used for making small holes, transplanting seedlings, mixing soil, and other small gardening tasks. Be sure to pick one with a comfortable grip and strong, rust-resistant materials.

- **Hose with a Watering Wand Attachment**: A flexible tube for transferring water from a spout to the plant, with a wand attachment that could extend reach and control water flow. It allows you to water multiple containers with a gentle spray, which is especially useful for hanging baskets or distant containers.
- **Hand Fork**: A small, handheld tool with several pointed prongs. It is used for loosening soil, weeding, and aerating around your plants, ensuring better water and nutrient penetration. Look for a hand fork with a comfortable grip and strong, rust-resistant materials.

Monitoring and Maintenance Tools

Keeping an eye on your plants' health and soil condition is crucial for a thriving garden. Some monitoring tools include:

- **Moisture Meter**: A device that measures moisture levels in the soil. It helps you determine if your plants are getting the right amount of water, preventing overwatering or underwatering. Insert the meter into the soil at different depths to get an accurate reading.
- **Soil pH Tester**: A device that measures the acidity or alkalinity of the soil. It allows you to check the soil pH and adjust it to suit the needs of your plants, ensuring optimal growing conditions. Test your soil pH periodically and adjust it using lime to raise the pH or sulfur to lower the pH as needed.

Protective Gear

Gardening can be physically demanding, so protecting yourself is just as important as caring for your plants. Here's what you need:

- **Gardening Gloves**: Protective gloves designed specifically for gardening. They protect your hands from thorns, sharp tools, and soil contaminants, preventing cuts, blisters, and infections. Wear gloves whenever handling soil or plants.

- **Wide-Brimmed Hat**: A hat with a large brim that provides shade and shields your face and neck from the sun, preventing sunburns and reducing the risk of heatstroke during long hours in the garden.
- **Sunscreen**: A lotion or spray that protects your skin from harmful long-term UV ray exposure.
- **Kneeling Pad or Knee Pads**: Cushioned pads protect your knees while working on the ground. Support for your knees while providing comfort, reducing pain during extended periods of planting or weeding. Use when working on the ground to minimize strain and avoid knee injuries.

Now that you're more familiar with the tools and understand why they're important, you can find the gardening tools and gear checklist on the next pages to keep track of the ones you may need when you're working on your own container garden.

Watering can with a large spout
Hand fork
Trowel
Scissors
Secateurs
Hose with a watering wand attachment

Gardening Tools and Gear Checklist

Use this checklist to ensure you have all the necessary tools and gear for your container gardening.

Basic Tools

- ☐ Watering can with a large spout
- ☐ Scissors
- ☐ Secateurs
- ☐ Trowel
- ☐ Hose with a watering wand
- ☐ Hand fork

Monitoring and Maintenance Tools

- ☐ Moisture Meter
- ☐ Pruning Shears
- ☐ Soil pH Tester

Protective Gear

- ☐ Gardening gloves
- ☐ Wide-brimmed hat
- ☐ Sunscreen
- ☐ Kneeling pad or knee pads

1.6 Beginner-Friendly Container Plants

When you're just starting out with your container gardening journey, it's better to care for plants that are easier to manage, rather than just jumping into ones that may have more requirements and have strict water scheduling.

We'll separate this section into two parts: full-sun-loving plants and shade-loving plants. Some plants prefer being under the sun for long hours, while others prefer the shade with very little sun exposure.

Sun-Loving Plants

Suppose you ever plan on setting up an outdoor container garden in your home or apartment, especially if you have a balcony, and the garden you plan on setting up doesn't exactly have enough shade. In that case, you can opt for plants that love basking in the sun and can tolerate its heat.

Choosing Sun-Tolerant Plants

Here are some easy-to-manage outdoor plants that can flourish under the sun, especially the more colorful ones that can brighten your garden:

- **Sunflowers**: Easy to grow from seeds, require minimal care once established.
- **Marigolds**: Hardy and resistant to pests, thrive with basic care.
- **Petunias**: Low-maintenance and prolific bloomers.
- **Zinnias**: Tolerant of various conditions, simple to cultivate.
- **Daisies**: Hardy and require minimal attention.

The next set of plants is still easier to grow, but they just need a bit more care in certain areas:

- **Geraniums**: Adaptable and need occasional watering and deadheading.
- **Lavender**: Drought-tolerant; just needs well-drained soil and sunshine.

If you don't want to *only* grow flowers, there are outdoor herbs and succulents that love to grow and receive full sun:

- **Rosemary**: Low-maintenance and drought-tolerant.
- **Sedum**: Very easy to care for, drought-tolerant succulents.
- **Yucca**: Requires minimal water and care.
- **Agave**: Thrives with little water and care.

Placement Strategies

Sun-loving plants need abundant sunlight, but proper container placement is crucial to ensure they receive the maximum benefit from the sun:

- **Optimal Positioning:** Place your containers in areas that receive at least 6-8 hours of direct sunlight each day. South-facing windows or open spaces in your garden are ideal.
- **Spacing:** Ensure enough space between containers to avoid shading each other, allowing each plant to get ample sunlight.
- **Elevated Placement:** Use plant stands or shelves to raise containers and maximize sun exposure, especially if ground-level spots are limited.

In case you want to grow this type of plant but your home lacks a bit of the sunlight needed, you can still supplement these plants with sunlight by incorporating reflective surfaces that can bounce light off of them:

- **Mirrors or reflective panels:** Position mirrors or reflective panels around your garden to bounce sunlight onto the plants, especially useful in areas with partial shade.
- **Light-colored walls:** Place containers near light-colored walls or fences to reflect additional light onto the plants.

Adjusting Care for Sunny Conditions

Sun-loving plants have specific care requirements to thrive under intense sunlight:

- **Watering:** These plants typically need more frequent watering due to higher evaporation rates. Ensure the soil remains consistently moist but not waterlogged.
- **Soil Quality:** Use well-draining soil to prevent water retention and root rot. Adding organic matter can enhance soil quality and support healthy growth.
- **Mulching:** Apply a layer of mulch on the soil surface to retain moisture, regulate temperature, and reduce weed growth.
- **Feeding:** Sun-loving plants often require more nutrients. Use a balanced, slow-release fertilizer to provide a steady supply of essential nutrients throughout the growing season.
- **Pruning:** Regularly prune to encourage bushier growth and remove any dead or damaged foliage.
- **Pest Management:** Keep an eye out for pests like aphids, whiteflies, and spider mites. Use organic pest control methods to keep these pests at bay.

These strategies will help your sun-loving plants flourish, providing vibrant blooms and healthy growth.

Shade-Loving Plants

These shade-loving plants are the perfect indoor plants for indoor container gardeners, especially if you don't have an accessible balcony or room with large windows that let in enough sunlight to pass through.

With the right approach, creating a thriving container garden in areas with limited sunlight is entirely possible. There are many shade-tolerant species that can flourish in low-light conditions.

Choosing Shade-Tolerant Plants

There are a variety of plants that thrive in shaded areas. The common shade-loving plants are usually the following:

- **Hostas:** Known for their lush foliage, they grow well in full to partial shade.
- **Ferns:** Various types adapt well to low light, adding texture and greenery.
- **Impatiens:** These flowers provide bright blooms even in significant shade.
- **Peace Lily:** Dramatic white flowers and dark leaves add a striking visual contrast to shaded spots.

If you want to diversify your garden further, these are options you can consider:

- **Caladiums:** With their colorful, heart-shaped leaves, caladiums thrive in shade and add a vibrant touch.
- **Begonias:** Ideal for shaded areas, begonias come in various types, including those with striking foliage and continuous blooms.
- **Astilbes:** These plants produce feathery plumes of flowers and do well in partial to full shade.

- **Heucheras (Coral Bells):** Known for their colorful foliage, heucheras can thrive in shaded conditions and provide year-round interest.

If you want some easily accessible vegetables and root crops that love the shade, there are a few kitchen friends you could grow as well:

- **Beets:** They love partially shaded areas and cool temperatures.
- **Kale:** Some variants of kale are bred to tolerate shaded areas; be sure to choose the correct variety!
- **Arugula:** A peppery and tarty veggie used for seasoning salads. This easily grows in the shade.

Placement Strategies

These plants love the shade, but it's best to learn how to properly place your containers strategically so they can still get enough sun for nutrients:

- **Rotational Placement:** Periodically rotate your containers to ensure all sides of the plant receive light, promoting uniform growth and preventing leaning.
- **Tiered Planting:** Arrange plants in layers, with taller plants behind shorter ones, allowing light to filter down and reach each plant according to its needs.
- **Reflective Surfaces:** Place reflective surfaces like white walls or light-colored tiles near your containers to amplify and reflect available light onto the plants.

Using Artificial Lighting

If you live in especially deeply shaded indoor areas, or if you have balconies that are overshadowed by buildings, then artificial lighting can supplement natural light.

One of the best types of artificial light for shade-loving plants is known as grow lights, which work as follows:

- **Use of Grow Lights:** These lights replicate the spectrum of sunlight, providing the necessary wavelengths for photosynthesis.
  - **LED Options:** LEDs are energy-efficient and emit less heat than traditional bulbs.
  - **Placement:** Position lights a few inches above the plants to mimic overhead sunlight, keeping them on for 12-14 hours daily. Remember, for healthy growth, shaded plants will need about 6-12 hours of darkness.
  - **Types:** Consider full-spectrum LED grow lights for the best results, as they provide a balanced light like natural sunlight.

Adjusting Care for Shaded Conditions

Shade-loving plants require a slightly different care routine, so here are some essential tips you need to know:

- **Watering:** Plants that grow in the shade need less watering since these plants have slower evaporation rates.
  - **NOTE:** Allow the soil to dry slightly between waterings to prevent root rot.
- **Soil Quality:** Make sure that the soil in your containers drains well and is rich in organic matter to support healthy root systems.
- **Humidity:** Some shade-loving plants, like ferns, thrive in higher humidity. Consider grouping plants or using a humidity tray to increase moisture levels around the plants.

- **Feeding:** These plants typically require less fertilizer. Use a balanced, slow-release fertilizer to provide a steady supply of nutrients without overwhelming them.
- **Pruning:** Regularly prune any leggy growth to maintain a compact shape and promote healthy development.
- **Pest Management:** Shade-tolerant plants can be more susceptible to pests like spider mites, aphids, and scale insects. Use organic pest control methods to keep these pests at bay.

With these tips in mind, once you've made the necessary changes and adjustments for your shade-loving plants, you've ultimately paved the way to a stronger foundation for your container garden's survivability.

Allowing yourself to gain more experience in gardening by starting with beginner-friendly container plants, you're not only jumpstarting your gardening hobby but also promoting a more learned and patient mindset when putting together a garden. That way, you're preparing yourself for more complex plants requiring special attention and time.

Chapter 2

Design and Planning

Now that you're prepared and ready for container gardening, you're also finally ready to decorate and plan your preferred aesthetic for your container garden.

Picture walking into a garden where every color, texture, and layout come together beautifully. That's what skillful garden design in container gardening is all about—using limited space to spark creativity.

In this chapter, you'll learn how to transform your garden into an efficiently functional, visually stunning, and comfortable space for you and your plants. We'll cover color coordination, texture variety, and height differences to help you create a garden that reflects your personal style and environment.

2.1 Designing with Color Coordination

Color coordination in garden design is about creating depth, mood, and visual pathways. By using color theory, you can significantly enhance your garden's appeal.

Color Theory and Color Schemes

To properly apply color theory to your garden, consider both the colors of blooms and foliage. For example, silvery-grey lavender leaves complement bright yellow marigolds, providing visual rest between bursts of color.

Take note of complementary colors, like blue and orange, for vibrant contrasts and analogous colors, such as green, blue-green, and blue, for a more harmonious and soothing look. Making thoughtful color choices enhances your garden's beauty and atmosphere.

When envisioning your container garden, consider it a canvas where colors play a crucial role in defining the space and setting the mood. Start by selecting a color palette, whether it's your favorite or if it's with intention, involves understanding the psychological effects of colors and how they transform the perception of space.

Some fundamental color palettes include:

- **Cool Colors:** Blues and greens create a sense of calm, perfect for a relaxing garden nook.
- **Warm Colors:** Reds and oranges energize the space and draw attention.
- **Analogous Colors:** Adjacent colors, or colors beside each other on the color wheel, like blues and violets, provide a sense of soothing uniformity.
- **Complementary Colors:** Colors that are directly across or opposite from each other on the color wheel, such as yellow and violet, add vibrancy and contrast.

If you prefer using one color instead of different colors, try going for a more monochromatic color scheme by using different shades of one color to give your garden a sophisticated and cohesive look. This approach simplifies plant selection and can easily unify the garden's aesthetic.

Below are some examples:

- **Purple Garden:** Combine lavender, purple coneflowers, and purple salvia for a serene and unified look.
- **White Garden:** Use white roses, white lilies, and white phlox for a clean and elegant theme.

Using accent colors is another way of designing your garden layout, which is the use of one or two accent colors to draw attention to specific areas or plants, acting as a 'feature' in your garden. You can easily achieve this by arranging a bright-colored flower close to dull or dark plants. Accent colors can highlight garden features or lead the eye through the space:

- **Red Accents:** Bright red poppies or geraniums can stand out against a backdrop of green shrubs.
- **Yellow Highlights:** Sunflowers or yellow daylilies can brighten up shady garden corners.

Creating contrast with accent colors isn't just for flowers; you can also use the colors of the foliage. Variegated leaves can add texture and color contrast even without blooms.

- **Silver Foliage:** Dusty Miller or lamb's ear can provide a soft, silvery contrast.
- **Purple Foliage:** Plants like heuchera or Japanese maple add depth with their rich hues.

Next, if you want variety in your garden year-round, you can try a more seasonal color scheme, where you start by planning for a succession of blooms throughout the seasons to keep your garden vibrant and fresh. To do this, you need to take note of which plants grow in each of the four seasons.

Here are some examples of flowers you could grow for the following seasons, along with seasonal color changes that could help ensure year-round interest for each season's unique palette:

- **Spring:** Soft pastels with tulips, daffodils, and cherry blossoms.
- **Summer:** Bold, bright hues from roses, daylilies, and hydrangeas.
- **Autumn:** Rich golds and reds with mums, asters, and autumn foliage.
- **Winter:** Evergreens or bright plants with colorful berries like holly or winterberry against white snow. You can also introduce plants with foliage that changes color during the fall for a smoother visual transition.

Finally, consider selecting colored pots and containers that complement or contrast your plants for added color coordination.

- **Bright Containers:** Use vibrant pots to add color accents in small spaces. These can be used with bright, same-colored plants for a unified look or with dull plants for a bold look.
- **Neutral Containers:** Terracotta or stone pots provide a classic look that allows any plant colors to shine. This works in any space, too.

Garden Arrangement and Planning

When laying out your container garden, consider the various containers or pot shapes, sizes, and materials while keeping your planned blooms or vegetables in mind.

Consider the following tips in choosing the right containers:

- **Matching Style:** Select containers that match the theme of your garden—rustic wooden containers for a cottage garden or sleek, modern pots for a contemporary look.
- **Size Variation:** Use a variety of container sizes to create visual interest while catering to different plant needs.
- **Groupings:** Arrange containers in groups of odd numbers for a more natural look. Elevate some containers using stands to add height variation.

When arranging your containers, you can try creating focal points in your garden. Having focal points guides the viewer's gaze and adds architectural interest, and these can be anything that directly attracts the attention of a viewer, such as:

- A brightly colored container surrounded by a dull or pale display
- An unusually shaped plant in an empty space or an arrangement of uniformly arranged plants
- A large, dramatic plant, like a Bird of Paradise, that towers over smaller blooms or plants.

Strategically place your focal points to lead the viewer's eye through the garden, highlight specific areas, and surround focal points with more subdued plants or colors to enhance their impact without causing visual clutter.

Utilizing accent plants, as mentioned in the previous sections, is another way to position focal points. Repeating specific plants throughout the garden provides unity and flow.

For example, using a silver foliage plant as an accent amidst deep greens and bright flowers can unify the color scheme and add elegance. You can also consider adding vertical elements like trellises or garden art to create height and additional focal points.

Lighting can also dramatically transform your garden, extending its usability into the evening and highlighting its beauty; you can use this as a way to create focal points as well. Consider these lighting strategies:

- **Path Lights:** Illuminate walkways to ensure safety and add a welcoming glow.
- **Spotlights:** Highlight focal points such as sculptures, water features, or standout plants
- **String Lights:** Drape string lights around pergolas, trees, or along railings to create a festive and cozy atmosphere.
- **Solar Lighting:** For an eco-friendly option, use solar-powered lights, which are easy to install and move.

Other than lighting and laying out your container garden effectively, another element that could set up a visually engaging garden is a good balance of colors, along with the various textures and hues of foliage.

Plant textures can range from the glossy leaves of a gardenia to the rough foliage of a fern. This interplay enhances the depth and complexity of your garden.

Some examples:

- **Combining Textures:** Arrange bright red flowers next to dark green, coarse-leaved plants to highlight colors and textures.
- **Achieving Harmony:** Experiment and adjust to balance elements, preventing any one texture or color from overwhelming the space.

2.2 Combining Textures and Heights

Beautifying your container garden isn't just about picking and choosing the right colors; it also involves combining different textures that can create depth and interest and experimenting with the height of your container garden's plants and pots.

The contrast between plant textures can make a monochromatic scheme dynamic, and paying attention to height can help guide the viewer's eye, ensuring the arrangement feels balanced and intentional.

Texture Variety

There are many textures you can try out when putting your container garden together. For instance, the fine, feathery leaves of ferns contrast beautifully with the broad, glossy leaves of hostas, creating visual appeal. Similarly, the spiky drama of ornamental grasses against the soft mounds of moss adds variety.

One way you can apply texture is by using plants with different textures to create borders or edges, also known as edging plants. For example, the feathery dill foliage can act as an edge for a path lined with the broad leaves of hostas, providing a textural contrast that enhances both plants. Here are a couple more examples:

- **Herb Borders:** Create borders with herbs like dill, fennel, or cilantro, which have fine, airy foliage.
- **Mixed Borders:** Combine coarse-textured plants like lamb's ear with fine-textured ones like maidenhair fern for a balanced look.

Another way is by implementing it on outdoor hardscape elements in the form of pathways. These have textures such as stone, gravel, and wood chips that can add texture and contrast to your overall garden aesthetic plantings.

These elements can define spaces within your garden and provide a different tactile experience.

To give you an idea of how this could work visually, imagine a pathway lined with smooth river stones that can contrast with the rugged texture of bark mulch in planting areas. Some examples include:

- **Gravel Pathways:** Use fine gravel to create a minimalist look that complements spiky plants like agave or yucca.
- **Wood Chips:** Ideal for natural, woodland-themed gardens, pairing well with ferns and hostas.

Texture isn't limited to plants alone; it extends to blooms and containers. Rough terra cotta pots can offset the delicate petals of roses, while sleek modern containers highlight the architectural lines of succulents.

Mixing textures creates visual layers that invite exploration and add complexity to your garden.

Height and Structure

Varying plant heights and structures add dimension and interest to your garden. Use tall plants like grasses or bamboo as focal points and layer shorter plants around them. Trellises and vertical gardening can maximize space and add visual intrigue.

For a beautiful cascading effect in your container garden, try arranging plants in tiers. This type of height arrangement is known as staggered planting, which can be done by placing taller plants like sunflowers at the back, mid-height plants like daisies in the middle, and low-growing plants like pansies at the front.

You can go a bit further by adding herbs below the pansies, making them not only *look* compelling but also functional. This arrangement also guarantees that each plant gets the sunlight it needs. Here are some examples:

- **Sunflower Backdrop:** Use sunflowers or hollyhocks at the back of your garden to provide height and structure.
- **Layered Effect:** Mid-height plants like lavender or coneflowers can fill the middle layer, with low-growing plants like creeping thyme at the front.

If you want to achieve a staggered look better, you can utilize multi-level planters or tiered plant stands to add height and visual interest without wasting extra floor space.

These are especially useful for small balconies or patios where space is a bit more limited. Some notable multi-level planters are:

- **Tiered Stands:** Use tiered stands for a mix of herbs and flowers, creating a compact but visually rich display.
- **Hanging Baskets:** Incorporate hanging baskets at different heights to add vertical layers, drawing the eye from the ground up.

Vertical Gardens

Now that you're familiar with designing and planning gardens on the ground, you're now ready for more complex layouts like vertical and hanging gardening.

The first complex layout you can do is the vertical container garden arrangement. This type of gardening allows you to expand your garden upwards, making the most of limited space, especially in urban environments.

Gardening structures like trellises, shelves, and tower gardens will allow you to grow a variety of plants without using too much floor space, which can make your garden area look more spacious and more organized.

- **Trellises:** These support climbing plants such as clematis, jasmine, and certain tomatoes and peas. Adds greenery and rustic charm to your garden while also providing privacy and shade.
    - **Material:** Wood, metal, or repurposed materials (e.g., ladders)
    - **Considerations:** These can break when overloaded with heavy plants, and may cause some plants to become stunted if the trellis is too restrictive.

   - **Set-Up**: Must be firm and able to withstand the weight of mature plants and potential wind loads. Regularly tie and train plants to the trellis for better upward growth and ensure stability.
- **Shelves and Racks**: These utilize limited space. Wall-mounted shelves or freestanding racks can hold multiple containers at different levels, creating a tiered garden effect.
   - **Set-Up**: Place near sunlight sources or supplement with growth lights that are enough for each plant. Arrange from tallest at the back to shortest in the front for easy and efficient sharing of light, air, and plant harvesting.
- **Tower Gardens**: Compact vertical system that is great for growing leafy greens and herbs while allowing a high yield.
   - **Set-Up:** Can be commercially purchased or DIY projects using PVC pipes, stacked pots, or repurposed materials, and often includes a central watering system that evenly distributes water and nutrients. These can also be moved around to capture optimal sunlight.
   - **Maintenance:** Regularly rotate the tower for even light exposure and check for pests.

Hanging Gardens

Next, hanging gardens add elegance and sophistication. Similarly to vertical gardens, these would expand vertically to maximize space more efficiently, just downwards from the ceiling or directly mounted on the walls.

- **Hanging Planters or Baskets**: Suspended from ceilings or brackets, these are perfect for trailing or cascading plants (e.g., begonias, orchids, black-eyed Susans, etc).

- **Materials**: Lightweight materials like plastic and resin to avoid overloading supports.
- **Set-Up:** These can dry out fast and will benefit more from a consistent watering schedule or self-watering system.

- **Wall Planters:** These turn bare walls into lush, living murals with herbs and flowers like basil, parsley, or petunias. Great for small spaces.
    - **Materials:** Fabric pockets, mounted pots, vertical frames with built-in pots.
    - **Set-Up**: Needs proper drainage to prevent water damage to walls and prevent root rot; consider a drip irrigation system to maintain consistent moisture levels. These tend to dry out quickly.

2.3 Companion Planting

Another way to optimize and enhance your plant's growth and health is through companion planting. This method involves placing compatible plants together where they can provide mutual benefits like pest control, improved pollination, and better resource sharing.

Benefits of Companion Planting

Companion planting offers extensive benefits, particularly in natural pest control. Certain plant combinations naturally repel specific pests without the need for chemical interventions. For instance, the strong scent of herbs like rosemary and thyme can confuse and repel pests attracted to vegetable plants.

Significant garden pollinators, such as bees and butterflies, are especially attracted by well-paired-up plants that could add a layer of protection against pests. These pollinators are crucial for the fruiting of many plants, and their presence is encouraged by the variety of plants and flowers that companion planting supports. This biodiversity leads to a more beautiful garden and promotes a healthier, more balanced ecosystem where pests are less likely to dominate.

Companion planting can also improve soil health. Plants like beans and peas are known for their nitrogen-fixing ability, enriching the soil for other plants. This mutual support system helps maintain nutrient levels in your containers, reducing the need for additional fertilizers.

Companion Planting Strategies

Strategically arranging plants and successfully cultivating them together as companions can maximize the output of companion planting. You can begin by understanding each plant's growth patterns and needs—some might require more sunlight, while others need more space to spread.

Arrange taller plants like tomatoes at the back of your containers to provide shade for more delicate, lower-growing plants like lettuce. Utilize vertical space by adding climbing plants like cucumbers on a trellis, creating a living green backdrop for smaller, bushier plants in the foreground. This maximizes space and creates microclimates within your containers where each plant can thrive.

One great example of a companion strategy, which you also must become familiar with, is the Three Sisters method. This method, originating from north of Mesoamerica, is a classic example involving corn, beans, and squash.

Corn could act as a natural pole for bean vines to climb, while beans could also adjust the nitrogen content in the soil, which benefits the three plants—and squash can spread along the ground, blocking weeds and retaining soil moisture.

This combination maximizes both vertical and horizontal space, creating a micro-ecosystem where each plant supports the others.

Another great companion planting strategy is mastering the integration of herbs among your vegetable containers to effectively utilize space and enhance your garden's health.

Certain herbs deter pests while also attracting beneficial insects to your garden:

- **Rosemary and Thyme**: These herbs deter pests with their strong scents, masking the smell of vegetables and making them harder for pests to locate. They are particularly effective when planted near brassicas like cabbage and broccoli.
- **Dill Near Cabbage or Onions**: Dill can naturally repel aphids and mites, which can significantly reduce the need for chemical pesticides. It also attracts ladybugs, hoverflies, and similar insects that prey on common garden pests.
- **Mint with Cabbages**: Planting mint near cabbages can help repel cabbage moths, ants, and aphids. However, mint is invasive and should be grown in its own container to prevent it from overtaking other plants.

Using flowers as plant companions for vegetable containers is another strategic way to increase productivity, enhance pollination, and manage pests while beautifying your garden.

Other great flower companions include:

- **Marigolds:** A vibrant colored flower that repels nematodes and other insect pests that aim for vegetable roots.
- **Nasturtiums:** A beautiful decoy plant that attracts aphids away from more valuable plants and draws in essential pollinators such as bees. Great companions for tomatoes and cucumbers

Common Companion Plant Pairings

There are many common plant pairings that are regularly used by gardeners all around. One common pairing is tomatoes and basil. Basil helps repel pests like mosquitoes and flies while potentially enhancing the flavor of tomatoes. The strong scent of basil confuses pests, making it harder for them to find the tomatoes. This pairing is beneficial in the garden and convenient for cooking, as these two plants are commonly used together in recipes.

Another good and commonly used pairing is marigolds and any vegetable—planting marigolds close to vegetables can deter nematodes and other pests, protecting plants like tomatoes and peppers. Marigolds release alpha-terthienyl, a substance that is toxic to nematodes and other soil pests, while its bright flowers attract insects such as ladybugs and parasitic wasps, which help control aphid populations.

Lastly, chives and carrots are another good pair. Chives repel carrot flies and enhance carrot growth with their pungent smell, which can confuse and deter carrot flies. Chives also help improve the flavor and growth of carrots.

Avoiding Bad Combinations

However, not all plants make good neighbors. Here are some combinations that you *must* avoid. Tomatoes and corn are two vegetables that shouldn't be planted together as they will be forced to compete for water and nutrients, and corn can overshadow tomatoes, blocking sunlight and stunting their growth.

Another inadvisable companion combination is fennel with some common plants like beans and tomatoes. Fennel releases allelopathic chemicals that inhibit the growth of many plants, so it's best to keep fennel isolated to prevent it from impacting the vitality of your other garden plants. However, if you want to grow fennel with a companion plant, here are some vegetables and plants to consider:

- Hydrangeas
- Chamomile
- Dill or cucumber
- Peas
- Mint
- Lettuce
- Common sage

2.4 Year-Round Planting

In section 2.1 of this chapter, we briefly talked about seasonal planting in consideration of the overall color and design of your container garden. However, there is more to it than just simply using it for aesthetic purposes.

Rotational Planting

Rotational planting is a technique that involves rotating plants based on their growth cycles and seasonal strengths, allowing for continuous harvests and sustained visual appeal. Rotational planting can help your garden remain vibrant and productive throughout the year.

To do this, start by dividing your container garden into sections and plan a rotation schedule; summer crops like tomatoes and basil fade when the season ends, so you replace them with cool-weather crops such as kale and Swiss chard.

This approach helps manage soil health by varying nutrient demands and discouraging pests from disrupting their cycles, while also allowing you to experiment with plant positioning, improving your garden's layout over time.

Soil Health Management

In addition to managing plant health, rotational planting is beneficial for maintaining soil health. Different plants have varying nutrient requirements; rotating them periodically ensures that no single nutrient is depleted from the soil. For instance, legumes like beans and peas add nitrogen to the soil, which can benefit leafy greens planted afterward. Practicing this reduces the need for and use of synthetic fertilizers and promotes a healthier, more sustainable garden ecosystem.

Selecting Seasonal Plants

Choosing the right plants for each season is crucial for a lively container garden year-round. Understanding the climatic needs of various plants and matching them to your region's seasons will allow you to have a better supply of vegetable crops year-round.

1. **Spring:** Plant cool-weather crops like lettuce, peas, and radishes. These thrive in mild temperatures and occasional frost.
2. **Summer:** Transition to heat-loving plants like tomatoes, peppers, and cucumbers. These need longer daylight hours and warmer temperatures.
3. **Autumn:** Grow fall favorites like pumpkins, squash, and kale, which tolerate cooler temperatures and shorter days.
4. **Winter:** Go for cold-hardy plants like evergreens, winter heather, and ornamental cabbage. For non-hardy plants, move containers to sheltered locations, use mulches, and consider protective covers like frost cloths.

Utilizing Microclimates

Maximize your garden's potential by using microclimates. These are small areas within your garden that have slightly different conditions from the general environment.

For example, a south-facing wall can provide extra warmth in winter, making it an ideal spot for cold-sensitive plants. Similarly, shaded areas can protect delicate plants from the scorching summer sun. Placing containers strategically can encourage favorable conditions for various plants year-round.

Transitioning Between Seasons

As the seasons change, so do your garden requirements. Summer plants may need more frequent watering and support, while winter plants require insulation to protect against the cold.

Adapting your gardening practices according to the season can help maintain plant health and ensure a productive garden.

Seasonal transitions often necessitate changes in your gardening routine to accommodate different weather conditions and plant needs.

Spring to Summer:

- **Increase Watering Frequency:** As temperatures rise, evaporation rates increase, necessitating more frequent watering. Ensure that your plants are adequately hydrated, particularly those in containers that dry out faster than garden beds.
- **Support Taller Plants:** As plants grow taller, they may require staking or caging to support their stems and prevent them from toppling over. This is especially important for plants like tomatoes and beans.
- **Pest Management:** With warmer weather, pests become more active, so take appropriate measures to control them by inspecting your plants regularly.

Summer to Autumn:

- **Gradually Reduce Watering:** Plants require less water as temperatures begin to drop. Gradually reducing the amount of water helps prepare plants for the cooler months ahead.
- **Begin Mulching:** Retain soil warmth and moisture by layering mulch around your plants. Mulch also suppresses weed growth and adds organic matter to the soil as it decomposes.
- **Harvesting and Pruning:** Harvest crops and prune back any dead or diseased foliage to prepare your plants for the dormant season while reducing the risk of disease.

Autumn to Winter:

- **Move Sensitive Plants:** Relocate sensitive plants to sheltered locations, such as a greenhouse or indoors, to protect them from frost and extreme cold.
- **Use Row Covers or Cloches:** For plants that remain outside, consider using row covers or cloches to provide an extra layer of protection against the cold. These can help extend the growing season and protect plants from frost damage.
- **Insulate Containers:** Wrap containers with bubble wrap or burlap to insulate the roots from freezing temperatures. This is particularly important for terracotta pots, which can crack in the cold.

Incorporating these strategies into your container gardening practice, you can ensure a productive and visually appealing garden throughout the year. Each season will bring its own set of plants and maintenance tasks, keeping your gardening experience dynamic and engaging.

2.5 Overwintering

Suppose you want to maintain and grow your plants during the winter. In this case, overwintering is essential to guarantee the survival of your perennial plants in their containers through the colder months.

Overwintering refers to methods used to protect plants from freezing temperatures and harsh winter conditions that can cause stress or even kill them. Many perennials, though hardy, may not survive freezing temperatures when their roots are confined to containers.

This is because container soil freezes more quickly and thaws more slowly, leading to root damage. Understanding effective overwintering strategies is crucial in safeguarding your plants' survival and thriving once spring arrives.

Cold-Hardy Plants

Selecting cold-hardy plants and protecting sensitive species is vital to maintaining a vibrant container garden during colder months. Here are plants that grow well during winter:

- **Evergreens**: Junipers, dwarf spruces, and hollies provide year-round greenery and structure.
- **Winter Heather**: Blooms in winter with vibrant colors, adding a splash of life to the cold season.
- **Ornamental Cabbages and Kales**: Decorative and cold-tolerant, these plants bring texture and color to winter gardens.

Bringing Plants Indoors

One effective way to protect your container plants during winter is by bringing them indoors.

However, transitioning plants outdoors to indoors requires careful planning to avoid shock. Follow these steps for a smooth transition:

1. **Container Material Considerations**: Terracotta and ceramic pots can crack in freezing temperatures. Consider using plastic or resin containers for plants that will remain outside all winter. If you're using ceramic pots, ensure they are frost-resistant or move them indoors.
2. **Gradual Acclimatization**: Start by slowly bringing the plants inside for a few hours each day over a few weeks during the fall, gradually increasing the time spent indoors. This helps them adapt to the changes in light, temperature, and humidity.

3. **Optimal Placement**: Once inside, place the plants in an area that best replicates their natural growing conditions. South-facing windows typically offer the most sunlight during winter months. If the area lacks natural light, place grow lights a few inches above the plants.
4. **Temperature Control**: Maintain consistent and appropriate temperatures. Most perennials prefer cooler temperatures during winter, mimicking their natural dormant period. Heat sources, like radiators and vents, should be kept away from plants to prevent drying out and disrupting dormancy.

Outdoor Overwintering

For plants that need to remain outdoors, here's what you can do to allow them to thrive:

- **Insulation**: Materials like straw, bubble wrap, or garden fleece can be used to insulate containers and protect roots from freezing temperatures. Wrapping these materials around the container in burlap can also provide additional protection.
- **Grouping**: Bundle containers together and place them in a sheltered location to reduce wind and cold exposure. The collective warmth can help moderate temperatures.
- **Mulching**: Layer thick mulch around the base of the plants to retain moisture and provide additional insulation. Organic materials like leaves or wood chips are excellent choices.
- **Elevating Containers:** Place containers on a platform or pot feet to prevent direct contact with the cold ground, reducing the risk of frost damage from below.

Extra Maintenance and Protection

Once you're done setting up your garden for overwintering, you need to start considering the following strategies to keep your plants healthy and warm during the winter season:

- **Shelter Locations**: Move containers to protected areas like south-facing walls, garages, or covered patios to reduce exposure to harsh elements and provide plants with warmth and passive solar heat.
- **Frost Cloths and Burlap**: Cover plants with frost cloths or burlap during notably colder nights to trap heat and prevent frost damage. These materials allow plants to breathe while providing thermal protection.
- **Small Greenhouses or Cold Frames**: Use these structures to prolong the growing season and protect delicate plants, creating a warmer microclimate. They are handy for starting seeds earlier in the season and protecting young plants from sudden cold snaps.
- **Watering Considerations**: Water plants well before they freeze, but reduce watering frequency during winter to avoid root rot. Moist soil holds heat better than dry soil.
- **Windbreaks**: Set up screen barriers to minimize wind exposure and cold damage. Plants can have an increased risk of dehydration and frostbite from cold winds.
- **Monitoring Weather Forecasts**: Keep an eye on your local weather forecasts to stay ahead of sudden cold snaps and be ready to quickly apply protective measures for your outdoor and indoor plants.
- **Using Heat Mats**: Heat mats can provide extra warmth at the root zone for particularly cold-sensitive plants, encouraging healthier growth during cold spells.
- **Monitoring Plants**: Check your plants for signs of stress or damage regularly. Adjust protection methods as needed based on the severity of winter conditions.

To give you a better idea of how you could apply the steps to your own container garden, here's an example of an overwinter setup:

1. Move potted evergreens and winter heathers to a sheltered spot—away from harsh winds while still able to receive enough sunlight.
2. Insulate soil with a 3-inch layer of straw mulch to maintain consistent soil temperature and prevent root damage.
3. Cover plants with frost cloths on cold nights in the late afternoon before temperatures drop and remove them in the morning when the sun warms the air.
4. Use a small greenhouse for delicate, non-frost-tolerant plants so they can survive the winter.
5. Reduce watering and keep the soil slightly moist to avoid waterlogged conditions. Check moisture levels regularly.

Reviving Plants After Winter

Once the winter season is over, reviving your plants afterward is a critical process that sets the stage for vibrant growth in the coming season:

1. **Gradual Reintroduction**: For plants brought indoors, reverse the fall acclimatization process by gradually reintroducing them to the outdoors as temperatures rise.
2. **Remove Protective Coverings**: For plants left outdoors, carefully remove any protective coverings and inspect them for signs of damage or disease.
3. **Pruning**: Trim any dead or damaged material to clean up the plant and encourage new growth.

4. **Soil Moisture**: Check soil moisture levels and rehydrate the soil slowly, being careful not to overwater, as cold soils retain moisture longer.
5. **Fertilization**: As new growth appears, start a fertilization regimen appropriate for each plant's needs to replenish nutrients depleted over the winter.
6. **Pest Inspection**: Check for any pests that may have overwintered in the mulch or soil and treat them accordingly to prevent early-season infestations and diseases.

2.6 Edible Flowers and Other Unique Plants

Besides beautiful and colorful blooms to visually improve your container garden, you can also grow other types of flowers and plants that are more exotic and may require some special extra steps of care, or grow some edible flowers to emphasize both the aesthetic appeal and culinary uses of your garden.

Caring for and Growing Edible Flowers

Most edible flowers prefer well-draining soil and sunny locations. Regularly watering them and providing organic fertilizers promote vibrant blooms, while deadheading, or removing spent flowers, encourages continuous flowering.

If possible, refrain from using harmful chemicals and settle on natural pest control methods, such as neem oil or insecticidal soap.

Some examples of edible flowers you can eat the leaves and petals of:

- **Nasturtiums:** Vibrant red, orange, and yellow blooms with a peppery taste. The leaves and petals are edible.

- **Pansies:** Mild, slightly grassy flavor that adds color to salads and soups. The leaves and petals are also edible.
- **Marigolds:** Bright flowers with a range of flavors, from citrusy to spicy. The petals are edible, but the leaves are less commonly used for cooking.

Unique and Exotic Plants

Adding unusual or interesting plants as either the main or supplementary focus can add to the overall appeal of your container garden:

- **Touch-Me-Not plant (Mimosa pudica):** responds to touch by folding its leaves downward almost bashfully, adding an interactive element to your garden.
- **Black Bat Flower (Tacca chantrieri):** striking black flowers that thrive in high humidity and low light conditions, ideal for shaded spots.

Combining Edible and Ornamental Plants

Pairing edible and ornamental plants in one section of your container garden can maximize utility and beauty.

For example:

- **Strawberries and Creeping Thyme:** Strawberries provide fruit, while thyme offers fragrant ground cover and culinary uses.
- **Nasturtiums and Passionflowers:** Nasturtiums can repel pests and be used in culinary dishes, while passionflowers attract pollinators such as bees and butterflies.

Inter-planting enhances visual appeal, deters pests, and improves pollination. Ensure compatibility by considering growth habits and requirements.

Chapter 3

Daily Care

We've covered the basic foundations and design strategies of container gardening, which means you're finally ready to learn more about your plants' daily maintenance and care.

This chapter contains daily practices that can help maintain your container garden's overall health while aiding in your growth to become a more competent container gardener.

3.1 Watering Techniques

Watering your plants can be as simple as pouring water over your pots to moisten their roots.

However, watering goes beyond a simple habit; it requires an in-depth understanding of each plant's unique hydration needs to make sure they receive the perfect amount of moisture for flourishing growth.

Watering Tools

When you're doing container gardening, water usage can get especially costly when you have multiple plants to take care of.

You can enhance the efficiency and effectiveness of watering with various tools:

- **Soaker Hoses:** Lay on the soil or beneath mulch, slowly releasing water directly to plant roots, reducing water wastage and keeping foliage dry to prevent diseases.
- **Drip Irrigation Systems:** A highly efficient way of directly transporting water to each plant via a network of thin tubing. This can be automated with timers.
- **Self-Watering Containers:** Feature reservoirs that allow plants to absorb moisture as needed, offering a worry-free solution.

Additional Tip: Use a watering can with a long spout for precise watering at the base of plants, minimizing water on leaves and reducing the risk of fungal infections.

Assessing Your Plants' Moisture Needs

Each species has its own unique water requirement that's shaped by its native habitat, root system, and stage of growth.

To monitor these plants' moisture levels and avoid over- or underwatering, you can do the following:

- **Moisture Meters:** Provides measurements of soil moisture content, indicating when watering is necessary.
- **Hands-On Methods:** Use a wooden dowel or finger to gauge soil moisture by feel and appearance. Insert the dowel into the soil; if the soil sticks, it indicates moisture levels.

Timing and Frequency

Your watering schedule and the number of times you water your plants can prevent multiple problems, such as the spread of plant diseases, overwatering, and root rot.

Here are the general rules for determining the plant watering schedule and the number of times you need to water them:

- **Early morning** allows plants to absorb moisture before the morning sun's heat increases the rate of evaporation.
- **Evening watering** is also feasible but should be done early enough to let the foliage dry before nightfall to reduce fungal infection risks.
- **DON'T water at noon**. The sun's heat will start evaporating the water fast, preventing your plants from absorbing the nutrients.

In terms of frequency, the number of times you should water your plants depends on the following factors: plant type, weather, and container type. Porous containers, like clay, dry quicker, while plastic pots retain moisture longer.

Adjust watering frequency based on these factors and the changing weather conditions.

Examples of watering frequency for plants:

- **Minimal Water:** Succulents, such as cacti and aloe vera, thrive in dry conditions, storing water in their fleshy leaves and needing minimal watering.
- **Medium Water:** Plants like lavender and rosemary prefer moderate moisture, needing a balanced approach to watering.
- **Lots of Water:** Moisture-loving plants like basil, ferns, and impatiens demand consistently damp soil to flourish.

Each plant has unique moisture needs, so make sure that you provide just the right amount without over- or under-watering. However, if you suspect that you may have overwatered or underwatered your plants, here's what you need to look out for:

- **Overwatered** plants have yellowing leaves, mushy stems, and root rot.
- **Underwatered** plants are wilting, with dry soil and browning leaf edges.

Besides water, heaters and dehumidifiers can affect your plants' moisture.

- **Heaters** can dry out the air, increasing the frequency of water needed.
- **Dehumidifiers** remove moisture from the air, potentially leading to faster soil drying.

3.2 Nutrient Management and Feeding

Water isn't the only source of nutrients for your container garden plants. Other sources of nutrients, such as organic fertilizer and adding minerals and vitamins, can benefit your plants' overall health.

Like other gardening techniques, managing nutrients in container gardening requires ongoing attention. Unlike traditional gardens, the confined space of a container means nutrients are consumed quickly and need to be replenished regularly to ensure healthy plant growth and productivity.

Essential Nutrients

Understanding the essential nutrients required for plant growth—nitrogen, phosphorus, and potassium—is critical to cultivating a thriving garden.

Here's what each essential nutrient does for the plant:

- **Nitrogen (N)** encourages healthy leaf and stem growth.
- **Phosphorus (P)** is crucial for root development and flower production.
- **Potassium (K)** enhances overall plant vigor and disease resistance.

It's important to remember these nutrients as they are necessary for knowing what your plants need. Here is a list of nutrient deficiency traits to watch out for, according to which nutrient the plant may be lacking:

- **Yellowing Leaves:** Indicates a lack of nitrogen.
- **Stunted Growth:** Points to phosphorus deficiency.
- **Brown Edges on Leaves:** Suggest a potassium shortfall.

Regular observation helps you address nutrient needs before they significantly impact plant health.

Organic Fertilization

Organic fertilization offers a sustainable and gentle way to enrich soil without harsh chemicals. Fertilizers like compost, plant-based materials, worm castings, animal manure, or liquid alternatives slowly release nutrients, improve soil structure, and encourage valuable microbial activity.

There is also a reduced risk of over-fertilization when using organic fertilizers. Regularly adding small amounts of these organic materials can help maintain nutrient levels.

When selecting organic fertilizers, consider the nutrient content, typically listed as N-P-K ratios on the packaging.

Different plants have varying nutrient needs at different growth stages, so choosing the right type and amount of fertilizer is crucial.

Examples of Organic Fertilizers:

- **Compost Tea:** A nutrient-rich liquid made by steeping compost in water. Ideal for regular feeding, it boosts plant health by providing readily available nutrients.
- **Application:** Soak tea in water for 24-48 hours to brew compost tea. Strain the liquid and apply it to the base of the plants every two weeks.
- **Fish Emulsion:** A quick-acting liquid fertilizer rich in nitrogen, making it excellent for leafy greens. It promotes vigorous growth and deep green foliage.

- **Application:** Dilute fish emulsion according to package instructions and apply it to the soil around the plants every 3-4 weeks.
- **Seaweed Extract:** Provides trace minerals and stimulates plant growth, enhancing resilience to stress and improving overall health.
- **Application:** Mix seaweed extract with water and spray on foliage or water the soil around the plants every 2-4 weeks.

If you want to create your own organic fertilizer, you can recycle kitchen waste.

Here's a simple recipe for your very own organic fertilizer:

- **Ingredients:** Coffee grounds (rich in nitrogen), eggshells (provide calcium and reduce soil acidity), and banana peels (offer potassium).
- **Preparation:** Dry the banana peels and eggshells thoroughly, then grind them together with the coffee grounds into a fine powder. Sprinkle this mixture around the base of your plants or mix it into the topsoil layer of your plants.

Timing Fertilizer Applications

Apply fertilizer every 2-4 weeks when plants are actively growing to maximize nutrient uptake and minimize waste. Avoiding the heat of the day by feeding your plants in the early morning or late afternoon can reduce plant stress and improve nutrient absorption.

You can also adjust the frequency of these feedings based on plant health and growth patterns. Over-fertilization can lead to nutrient burn, evident by brown, crispy leaf edges.

Managing Soil pH

The pH level of your soil affects plant health by influencing nutrient absorption. Most plants grow better in soil with a pH between 6.0 and 7.0, while some, like blueberries, require more acidic conditions. It's advisable to test your soil's pH periodically by using simple test kits from your local garden store.

Conduct soil tests to monitor nutrient levels and pH, adjusting as needed. You can do this by using a home soil test kit to check for pH, nitrogen, phosphorus, and potassium levels.

After determining your soil's pH levels, you can use soil conditioners to adjust the acidity levels accordingly. Adjusting pH levels can take several months to take effect fully. Two of the most commonly used soil conditioners are:

- **Sulfur:** A soil acidifier. This lowers pH, making the soil more acidic.
    - **Application:** Apply sulfur to the soil according to the recommendations on the product label, usually once a year.
- **Lime:** A soil amendment. This raises the pH, making the soil more alkaline.
    - **Application:** Apply lime in the fall so it has time to react with the soil before the growing season. Follow the package directions for the correct amount.

After you've determined your plants' nutrient levels and needs, you're set to tailor your plants' feeding schedule.

You can set this up according to the growth stages of your plants:

- **Seedlings and Young Plants:** Benefit from higher phosphorus content to encourage root development.
- **Mature Plants:** May require more nitrogen to support leaf and stem growth, especially leafy greens.
- **Flowering and Fruiting Plants:** Often need a balanced approach with a slight increase in potassium.

Every now and then, it's highly beneficial to rotate your plants. Changing the types of plants grown in each container can prevent nutrient depletion and reduce pest buildup.

For example, after growing heavy feeders like tomatoes, plant legumes such as beans or peas that fix nitrogen in the soil.

3.3 Pruning and Deadheading

Pruning is often seen as a chore, yet for container gardens, it's vital for maintaining plant health, improving their aesthetic form, and enhancing productivity.

In confined spaces of containers, plants can quickly become overgrown due to excessive watering and fertilizing, leading to poor air circulation and reduced light, making them vulnerable to diseases and pests.

Pruning on a regular basis helps maintain plant growth, promotes bushier foliage, and encourages the development of new branches, leaves, and flowers. This also improves structural integrity, which can help plants support their foliage and fruit better during harsh weather or when moving containers.

Proper Pruning and Deadheading Techniques

Correct pruning is important to avoid injuring the plant and opening the door to disease or pests. Here are some guidelines on how to start pruning your plants properly:

- **Select the Right Tool:** Use sharp secateurs, pruning shears, or a garden knife, depending on the thickness of the stems.
    - Make sure that your tools are properly sanitized and sharp to make clean cuts that heal quickly while preventing the spread of disease between plants.
- **Remove Dead or Diseased Branches:** Cut back to healthy wood, ideally just above a leaf node or bud, to encourage new growth.
- **Shape and Thin Out:** Selectively thin out crowded areas to reduce the plant and achieve the desired shape for better light penetration and air circulation.
    - Cut at a 45° angle to shed water away and reduce the risk of disease.
- **Timing:** Prune during the plant's dormant period or early in the growing season to minimize stress by identifying growth patterns (e.g., whether they bloom on new or old wood) to encourage maximum flowering.

Deadheading, the removal of spent flowers, is just as important for maintaining plant health and aesthetics as pruning. It prevents plants from diverting energy into seed production, encouraging more blooms instead.

However, it should be noted that not all flowers may need this, so it's best to check if your blooms can self-clean by shedding their spent flowerheads.

Some important reminders when deadheading plants:

- **Regular Inspection:** Check your non-self-cleaning plants regularly for spent blooms and remove them promptly to keep your garden looking tidy and to encourage continuous flowering.
- **Pinch vs. Cut:** For delicate flowers, pinching off spent blooms with your fingers can be more effective and gentler than using tools.

Pruning and deadheading play critical roles in disease prevention. Doing this properly can further increase your plant's lifespan.

Benefits include:

- **Improve Air Circulation:** Maintain an open, airy structure to reduce humidity and prevent fungal diseases like powdery mildew and leaf spot.
- **Remove Infected Blooms and Foliage:** Prevent the spread of diseases by promptly removing dead or infected parts.
- **Clean Up Debris:** Regularly clear fallen debris and spent flowers to minimize the risk of pests and diseases that lurk in decaying plant material.

Adding these techniques to your daily garden routine can be extremely beneficial, as each careful snip shapes the plant's growth and guides its future development, which keeps your garden dynamic and thriving.

3.4 Managing Pests and Diseases

Taking care of your container garden within your apartment isn't just about nurturing and maintaining your plants; it's about preventing pests and diseases from spreading and treating your infected plants. Addressing these issues organically will safeguard your plants' well-being while supporting the environment and your own health. The first step in maintaining a healthy and beautiful garden is understanding the common threats it may face.

Common Pests and Pest Control

Plant pests can be a hassle to deal with, and there are many to keep track of. However, in container gardening, there are a couple of common pests that you need to pay attention to:

- Aphids are tiny, often green insects that cluster on the undersides of leaves and stems, sapping the plant of their nutrients and leaving behind sticky honeydew, which can attract sooty mold.
- Spider Mites are even smaller and more insidious, causing leaves to look speckled with yellow by sucking sap from plants.
- Whiteflies resemble tiny white moths and swarm around the plant when disturbed, causing damage and spreading diseases.

To control the pests that harm your plants, try introducing these natural pesticides:

- **Natural Predators**: Introduce ladybugs or lacewings to your container garden. These feed on aphids, mites, and other pests, naturally reducing their populations.
- **Homemade Sprays or Soaps**: Use mild and pure natural soap solutions or garlic spray. Blend two bulbs of garlic with water, dilute in a gallon (3.785 L) of water, and spray on foliage.
- **Neem Oil**: Neem oil is a potent organic solution for many pests. It acts as an antifeedant and disrupts the life cycle of pests at any stage. Mix neem oil with water and a few drops of dish soap and spray it on the plants.

- **Insecticidal Soap**: A practical and organic way to control soft-bodied pests. This breaks down the pest's outer layer, leading to dehydration and death.

Common Diseases and Management

Pests aren't the only problem that you'll face when protecting your plants. There are also diseases that you have to take extra care of when one of your container gardening plants has them:

- Powdery Mildew is the unsightly white or gray splotches on leaves and stems, which may cause them to turn yellow and fall off. Young foliage is especially susceptible to this.
- Blight is a more serious case that can cause severe yellowing. Brown patches can be found on leaves and often require prompt intervention as they can suddenly appear in several different forms.

As long as you catch your plants having these symptoms, there are organic solutions that can help you treat them:

- **Prevention**: Ensure good air circulation by not overcrowding plants and pruning regularly. Remove affected leaves immediately to prevent spread.
- **Natural Fungicides**: Neem oil or a baking soda and water mixture can act as natural fungicides. Test any homemade spray on a small area first to ensure it doesn't damage foliage.
- **Copper Fungicides**: Copper-based fungicides are effective against a wide range of fungal diseases. They are considered safe for organic gardening when used sparingly. However, this works best as a deterrent and may not always be effective against the later stages of blight.

- **Sulfur**: Sulfur is another natural element that can be used to treat and prevent further fungal infections. It is particularly effective against powdery mildew, rust, and fire blight.

Integrated Pest Management (IPM)

Integrated Pest Management (IPM) may sound intimidating at first, but it's simply the holistic approach to managing pests and diseases in your garden. Under the IPM are four different methods for long-term prevention, namely, cultural, mechanical, biological, and chemical methods.

Cultural control involves practices that reduce pest establishment, reproduction, and survival. These practices create an environment that is less conducive to pest problems.

- **Crop Rotation**: Rotate plants to prevent pests and diseases from becoming established. For example, rotating tomatoes with beans can reduce soil-borne diseases and pests that target specific crops.
- **Sanitation**: Remove and dispose of diseased plant material to prevent the spread of pathogens. Clean garden tools and containers regularly.
- **Proper Plant Spacing**: Adequate plant spacing improves the overall air circulation of your container garden, which minimizes humidity and lowers the likelihood of fungal diseases. For instance, spacing tomato plants 18-24 inches (45.72-60.96 cm) apart helps prevent blight.

Mechanical control includes physical methods to remove or block pests from plants.

- **Handpicking**: Regularly inspect plants and manually remove pests like caterpillars and beetles. This method is effective for small infestations.

- **Barriers**: Set up physical barriers such as row covers or netting to shield plants from insect pests. For example, use floating row covers to protect cabbages from cabbage moths.
- **Traps**: Set up traps to monitor and control pest populations. Yellow sticky traps can be used to catch whiteflies and aphids.

Biological control uses natural predators, parasites, or pathogens to control pest populations.

- **Beneficial Insects**: Attract predatory insects, such as ladybugs, which prey on aphids, or predatory mites that target spider mites.
- **Nematodes**: Use beneficial nematodes to control soil-dwelling pests like grubs and root weevils.
- **Microbial Insecticides**: Employ microbial insecticides, such as Bacillus thuringiensis (Bt), to target specific pests like caterpillars without harming beneficial insects.

Chemical control involves using pesticides as a last resort and choosing the least harmful products to humans, animals, and beneficial insects.

- **Organic Pesticides**: Use organic options such as neem oil, insecticidal soap, or pyrethrin. Neem oil can control aphids, whiteflies, and spider mites.
- **Selective Application**: Apply chemicals selectively and only when necessary, targeting specific pests and avoiding broad-spectrum pesticides that can harm beneficial insects.
- **Proper Timing**: Apply pesticides when beneficial insects are less active, such as early morning or late evening.

Example of IPM in Action

Now that you've learned the four different IPMs, here are a few ways you could implement these methods:

1. **Cultural**: Rotate crops to disrupt pest life cycles and use resistant plant varieties.
2. **Mechanical**: Use row covers to protect young plants and handpick pests.
3. **Biological**: Introduce ladybugs to control aphids in the early morning or evening, and use the bacteria Bt (Bacillus thuringiensis) in the early morning or late afternoon to manage caterpillar infestations.
4. **Chemical**: If necessary, apply neem oil to affected plants, ensuring minimal impact on beneficial insects.

Here are some everyday IPM practices that must be observed to ensure your pest management's effectiveness:

- **Regular Monitoring**: Inspect plants frequently to identify pest problems early.
- **Record Keeping**: Keep a garden journal to track pest and disease occurrences, the effectiveness of treatments, and seasonal patterns. This helps in predicting and preventing future issues.
- **Education**: Stay informed about the latest IPM techniques and pest control methods.

By incorporating these IPM strategies and organic methods, you can effectively eliminate and prevent pests and diseases in your garden while promoting a healthy and sustainable environment.

Chapter 4

Eco-Friendly Practices

For many gardeners, container gardening is a means for a more self-sufficient and sustainable lifestyle. Growing produce and having a productive hobby are just some of the many benefits that come with it.

However, like many hobbies and lifestyle activities, it's important to act on earth-friendly practices that are not only beneficial for us, but also for our environment.

For this chapter, we'll focus more on discussing the important eco-friendly techniques and practices that can be used to promote a healthier garden while actively making a positive impact to reduce our personal environmental footprint.

4.1 Composting

Composting offers two rewards: it enriches your container garden soil and is one of the best methods for reducing waste.

Organic waste makes up a significant portion of landfill content, decomposing anaerobically and releasing methane, a potent greenhouse gas.

By composting kitchen scraps, leaves, and other organic waste, you divert this waste from landfills and transform it into compost, a nutrient-rich material that greatly benefits garden soil.

Composting enhances soil structure and improves water retention, which is crucial for container gardening where soil volume is limited. It also introduces beneficial microorganisms that help suppress plant diseases, transforming the soil into a living, breathing ecosystem.

Composting Methods Suitable for Small Spaces

Some efficient composting methods are achievable even in the smallest spaces. Unique techniques such as bokashi, worm composting, or even using simple countertop composters can fit into nearly any nook, making them ideal for urban apartments or homes with limited outdoor space.

For the two unique composting methods mentioned, here are some brief descriptions of how each of them works:

- **Bokashi Composting**: Uses a mix of microorganisms to ferment kitchen waste, including dairy and meat, in sealed containers, minimizing odors and speeding up the process.
- **Worm Composting (Vermicomposting)**: Uses red wrigglers to break down food scraps and paper into high-quality compost in a small bin, which is perfect for under the kitchen sink or on a balcony.

These methods efficiently manage waste and turn it into valuable compost, perfect for enriching container plants.

Using Compost in Containers

Compost is great for container gardening, but requires a thoughtful approach to avoid nutrient burn, where excessive nutrients can harm plant roots.

To prevent harming your plant roots, make sure the compost is mature and fully decomposed. Mixing one part compost and two parts potting soil is the proper way to use it. This gives your plants all the benefits of compost without the risk of nutrient overload.

After adding compost, regularly check on your plants to gauge their health and the compost's effectiveness, adjusting your approach as necessary.

DIY Compost Recipes

Creating your own compost blend tailored for container gardening enhances the autonomy and satisfaction of managing your garden's ecosystem.

A balanced compost recipe might include:

- **Green Material**: Kitchen scraps and fresh plant clippings.
- **Brown Material**: Dried leaves, twigs, or shredded paper.

Layer green and brown materials in a bin or tumbler by alternating between them. Maintain moisture by occasionally watering the pile and turning it regularly to introduce oxygen, accelerating decomposition. This homemade compost feeds your garden and opens the loop of your household waste by turning scrap into a resource.

To ensure that your compost can successfully provide the beneficial nutrients needed for your plants, these are what you need to consider:

- **Temperature Control**: Keep your compost pile warm, ideally between 135°F and 160°F (57–71°C), to speed up decomposition and kill off weed seeds and pathogens. If you live in colder climates or regions, insulating your compost bin or placing it in a sunny spot would be ideal.
- **Odor Management**: A well-balanced compost pile should not produce foul odors. Ensure a proper mix of green and brown materials to avoid excess nitrogen, which can cause smells. If your compost smells bad, add more brown material and turn it to aerate.
- **Pest Prevention**: If you want to keep pests out of your compost, start by avoiding meat, dairy, and oily foods. Secure lids on bins and use fine mesh to cover air holes if necessary.

4.2 Water Conservation

With regular watering comes water conservation. To practice eco-friendly gardening, it's best to reflect on environmental stewardship and practical wisdom.

By implementing efficient and resourceful water-saving techniques, such as using rain barrels, setting up drip irrigation systems, applying organic mulches, and adopting water-saving habits, you can sustain your garden while contributing positively to the broader ecosystem.

These methods ensure that every drop of water serves a purpose, enhancing the health and productivity of your container garden while minimizing waste and saving your water bill.

Rain Barrels

Capturing and using rainwater with rain barrels is an effective and straightforward method to reduce dependence on public water systems. This is especially useful in areas with water scarcity.

Rain barrels are available in local home and garden supply stores, but you can also check online marketplaces for sellers or wait for end-of-season sales to get discount barrels. You can also use unused barrels at home if they are available.

Once you have a barrel for rainwater, you can go ahead and securely position and connect your rain barrels to your home's downspouts to collect excess runoff water.

After that, there are a couple of factors to consider when you're keeping and maintaining rain barrels for your container garden:

- **Mosquito Prevention**: Use secure lids, screens, or larvicides to prevent mosquito breeding. Regularly check for any openings that might allow mosquitoes to enter.
- **Local Regulations**: Check local guidelines on rainwater harvesting to ensure compliance. Some regions have specific rules about the collection and use of rainwater.

Rain barrels supply you with a ready source of water and help manage stormwater runoff, reducing the burden on local water treatment facilities.

IMPORTANT: Rain barrels need to be used, maintained, and cleaned regularly to prevent the spread of mosquito larvae and algae. If stored correctly, rain barrels can store water for up to a week.

Drip Irrigation for Containers

Drip irrigation systems provide targeted watering that is both water-efficient and conducive to plant health. This method is particularly beneficial in container gardens where precise moisture and water control are crucial.

This system can go hand-in-hand with the use of rain barrels; once you have collected rainwater with your rain barrels, you can either transfer that water straight to the container that is connected to your plants through drip irrigation or you can connect your rain barrel directly to the drip irrigation system to provide a continuous water supply.

Drip irrigation containers come with some benefits:

- **Efficiency**: These deliver water directly to the base of each plant, minimizing evaporation and runoff.
- **Setup**: Modern kits are user-friendly and can be automated with timers for consistent watering. Drip irrigation systems can be easily modified to fit various container sizes and layouts.
- **Health Benefits**: Keeps foliage dry, reducing the risk of plant diseases that thrive in moist conditions. This direct watering method also ensures that water reaches the roots more effectively.

Water goes exactly where it's needed with drip irrigation, promoting healthier and more resilient plants.

Mulching to Reduce Evaporation

Using organic mulches in container gardening is a simple yet effective strategy for conserving soil moisture.

- **Materials**: Use shredded bark, straw, or leaf litter. Other options include compost, newspaper, and grass clippings.
- **Benefits**: Reduces evaporation, adds nutrients, and improves soil structure as it decomposes. Mulch also helps suppress weeds.
- **Application**: Apply a layer of at least 2 inches (5.08 cm), but keep mulch away from plant stems to prevent rot. Reapply mulch as needed to maintain its effectiveness and appearance.

Regularly replenish mulch to maintain its moisture retention and keep your garden looking tidy.

Water-Saving Habits

Adopting water-saving habits is crucial for sustainable gardening:

- **Timing**: Water early in the morning to allow water to soak deeply before evaporation accelerates. Early watering also helps plants cope better with the day's heat.
- **Method**: Use a watering can or hose attachment that delivers water slowly to minimize runoff. Soaker hoses and drip systems are also efficient methods.
- **Plant Needs**: Understand your plants' specific water requirements to avoid overwatering. Group plants with similar water needs together to make watering more efficient.

Implementing these habits requires mindfulness and observation, ensuring each plant receives the optimal amount of water without waste.

Advanced Water Conservation Techniques

Consider incorporating advanced techniques and tools to optimize water usage in your container garden:

- **Self-Watering Containers**: These containers have built-in reservoirs that provide consistent moisture to plants, reducing watering frequency.
- **Rain Gardens**: If you have some outdoor space, creating a rain garden can help capture and utilize runoff water effectively.
- **Greywater Systems**: Reusing household greywater (from baths, sinks, washing machines) for garden irrigation can significantly reduce water waste. Ensure that the greywater system is safe for plants and follows local regulations.

4.3 Heirloom Varieties

Seeds that have been passed down through generations are heirloom varieties, typically for over fifty years or more. These plants are known for their unique flavors, vibrant colors, and fascinating history.

Unlike hybrid plants, heirlooms are prized for their genetic diversity and ability to preserve horticultural history.

Choosing to grow heirlooms helps preserve genetic diversity and ensures these unique strains continue to thrive for future generations.

Finding Heirloom Varieties

Heirloom varieties can be sourced from various places, depending on where you live:

- **Seed Companies**: These companies specialize in preserving heirloom seeds. You could consider starting your search for high-quality seeds with reputable seed companies such as:
    - Southern Exposure Seed Exchange
    - Seed Savers Exchange
    - Baker Creek Heirloom Seeds.
- **Seed Catalogs:** These often contain websites or numbers for you to use to check for available stock. You can often find catalogs from stores around your area, such as:
    - Garden nurseries
    - Bookstores and libraries
    - Farmers' markets
    - Home improvement stores

- **Local Nurseries**: Some local nurseries carry heirloom plants and seeds.
- **Farmers' Markets**: Local farmers often sell heirloom seedlings and produce.
- **Seed Swaps**: Participate in community seed swaps to exchange heirloom seeds with other gardeners.

Selecting Heirloom Varieties for Containers

When selecting heirloom varieties for container gardening, consider size, growth habits, and climate adaptability. Taking note of these elements when choosing the right heirloom variety can be beneficial for your overall garden, as it can save time, space, and resources. Here are some recommendations for container gardening with heirloom varieties:

- **Size and Growth Habit**: Dwarf varieties of tomatoes or compact types of beans are ideal for containers due to their limited growth habit, which prevents them from becoming too large for confined spaces. Here are some compact tomato and bean varieties you can try:
    - 'Patio' tomato
    - Sungold tomato
    - 'Bush Blue Lake' beans
    - Dwarf ‘Amethyst’ French beans
- **Climate Adaptability**: Heirloom plants are known to be hardy. However, some heirlooms, having been cultivated in specific regional climates, might not perform well outside their usual environment. It’s still best to research heirloom varieties that thrive in your local conditions to bring an extraordinary range of flavors and colors to your garden. Examples include:
    - 'Brandywine' tomatoes for warm climates
    - 'Dragon's Tongue' beans for cooler areas.

Seed Saving for Sustainability

Many farmers and gardeners save heirloom seeds to re-grow the same variety every year. Seed saving is crucial for growing heirloom plants as this promotes sustainability and self-reliance. By saving seeds, you ensure a personal supply for the next planting season and contribute to preserving valuable genetic resources. Follow these steps to save seeds effectively:

1. **Selecting the Right Plants**: Choose the healthiest, most vigorous plants to harvest seeds from, and look for plants that have shown resilience and produced high-quality fruits or flowers throughout the growing season.
2. **Harvesting Seeds:**
    - **Fruits and Vegetables**: Allow fruits to ripen fully on the plant. This ensures that the seeds have reached peak viability. For example, tomatoes should be left to overripen on the vine.
    - **Flowers**: Let the flowers go to seed. This means allowing the flower heads to dry out on the plant until they can be easily crumbled to release the seeds.
3. **Extracting and Cleaning Seeds:**
    - **Fruits**: Scoop out the seeds and separate them from the pulp. For example, scoop seeds from tomatoes and rinse them thoroughly to remove any remaining pulp.
    - **Flowers**: Shake dried flowerheads, such as marigolds or sunflowers, to release the seeds.
    - **Cleaning**: Clean the seeds thoroughly to remove clinging fruit or plant material. This helps prevent mold and ensures that the seeds store well.

4. **Drying Seeds**: Spread the cleaned seeds on a paper towel in a well-ventilated area away from direct sunlight. Allow them to dry completely over several days. Proper drying is essential to prevent mold growth during storage.
5. **Storing Seeds**: Once the seeds are completely dry, place them in labeled, airtight containers such as envelopes, jars, or plastic bags. Store the containers in a cool, dark location to maintain seed viability. Labeling each container with the plant variety and date of collection helps keep track of your seed inventory.

Preserving Flavor and Nutritional Value

Heirloom varieties are often chosen for their superior flavor and nutritional profiles. Unlike many commercially grown hybrids, heirlooms are not bred for long-distance shipping or uniform appearance.

This means they can offer more robust flavors and higher nutritional content.

- **Flavor**: Varieties like 'Golden Jubilee' tomatoes offer a sweet, mild flavor, while 'Black Krim' tomatoes are known for their intense, smoky taste.
- **Nutrition**: Heirlooms often have higher levels of vitamins and antioxidants compared to hybrids. For example, 'Red Russian' kale is rich in vitamins A, C, and K, making it a nutritious choice for your garden.

By growing heirloom varieties, you can enhance your garden's biodiversity and enjoy the superior taste and nutritional benefits these plants offer.

Heirlooms like Purple Cherokee tomatoes, Sugar Snap peas, and Scarlet Runner beans are just a few examples of plants that provide unique flavors and rich nutritional profiles.

4.4 Wildlife-Friendly Gardens

Creating a wildlife-friendly container garden transforms your space into a lively intersection where human cultivation meets the natural ecosystem.

This approach enhances your garden's aesthetic and ecological value while supporting local biodiversity, creating a sanctuary for pollinators, birds, and beneficial insects.

Plant Choices for Supporting Wildlife

Before you start choosing specific plants to attract and support wildlife, these are plant types you need to consider.

- **Native Plants:** These plants need to be adapted to the local climate. They require less water and care and are more attractive to local wildlife. Here are some notable native plants for any region:
    - Black-eyed Susan
    - Purple coneflower
    - Wild bergamot
- **Food Sources**: Wildlife will need a good variety of food sources that can cater to their needs.
    - **Nectar and Pollen:** Flowers such as echinacea (coneflowers) provide seeds for birds, while milkweed provides nectar for butterflies.
    - **Fruits and Seeds:** Plants like serviceberry and elderberry provide fruits for birds and other wildlife.

Becoming more well-informed about these plant types, which benefit the animals and insects in your area, can be rewarding as it will also contribute to the local ecology.

Once you are more knowledgeable of the native plants and food sources of the wildlife in your area, the next step is to strategically prepare your container garden through diverse planting.

- **Seasonal Blooms:** Include a variety of plants that bloom at different times of the year to provide a continuous food source for wildlife.
- **Layered Planting:** Use a mix of trees, shrubs, and groundcovers to create diverse habitats.

Attracting Beneficial Wildlife

To attract wildlife, your garden should mimic the inviting essence of natural habitats. You will need to pay special attention to researching the native flora and fauna you wish to add and attract to your garden—it's always best to get to know the flowers and plants that grow around your area to prevent potential invasive species from outcompeting the local flora.

- **Plants for Pollinators:**
  - **Lavender and salvia:** Excellent for attracting honeybees with their abundant nectar.
  - **Butterfly bushes and milkweed:** Draw butterflies, providing them with nectar and supporting their lifecycle. Milkweed is particularly crucial for monarch butterflies.
- **Plants for Birds:**
  - **Sunflowers and coneflowers** attract birds, such as bluejays and finches, which help control pests by eating garden insects and providing seeds for them to eat.
  - **Berry-producing plants,** such as holly, elderberry, and serviceberry, offer food for birds throughout the year.

- **Plants for Beneficial Insects:**
  - **Yarrow and dill:** Attract ladybugs and parasitic wasps that prey on garden pests.
  - **Goldenrod and asters:** Support a variety of beneficial insects, including bees and butterflies.

Water Sources and Shelter

Planting food and native plants is not the only need that your local wildlife would need. Providing and properly maintaining water sources and shelter is vital for creating a welcoming environment for them.

- **Water Sources:**
  - **Birdbath:** A shallow dish filled with stones and clean, shallow water can quench the thirst of small birds and insects.
  - **Water Features:** Small ponds or fountains can attract a wide range of wildlife (e.g., frogs, toads)
- **Shelter:**
  - **Birdhouses:** Install birdhouses to offer nesting spots for various bird species.
  - **Wild Areas:** Leave some areas wild with dense shrubs or piles of rocks and twigs. These habitats offer nesting and hiding spots for various creatures.
  - **Bee Hotels:** Provide habitats for solitary bees, which are excellent pollinators.
- **Winter Shelters:**
  - **Leaf Litter:** Maintain a layer of leaf litter and leave some plants unpruned to offer shelter and food for insects during winter.
  - **Perennial Grasses:** Leave uncut to provide habitat and food for wildlife through the winter months.

When providing water and shelter for the local fauna, it's important to keep these necessary safety measures in mind to keep yourself and the animals you want to attract protected:

- **Stagnant Water:** Leaving water in open spaces for a long time, to the point it becomes standing water, can turn into a troublesome hotbed for mosquitoes, all of which may carry diseases. Make sure to keep water dishes fresh and clean, and water features maintained.
- **Predators:** When attracting small wildlife (e.g., hares, mice, other rodents), these may inadvertently bring in predators. Be mindful of creating safe havens for smaller creatures to protect them.
- **Chemical Usage:** Avoid using pesticides, herbicides, or other chemicals around shelters and plants, as these can be harmful to wildlife. Always go for natural alternatives whenever possible.

Observing and Enjoying Wildlife

By creating a space that welcomes and supports local fauna, you contribute to biodiversity and gain the opportunity to observe the intricate interactions within nature.

Watching birds, bees, butterflies, and other creatures interact with your garden can be a rewarding experience, providing insights into the delicate balance of the ecosystem and the importance of conservation efforts.

However, it's important to remember to be cautious and considerate when interacting with wildlife, as constant human interaction can be stressful for animals. Observe from a good distance, avoid touching or being overly vocal towards wild animals, and be kind to nature.

Share Your Experience—Make a Difference

Plant the Seeds of Inspiration

"To plant a garden is to believe in tomorrow." - Audrey Hepburn

Your kindness can inspire others. Let's spread the joy of gardening together!

Would you help someone else, just like you—wanting to grow a garden but unsure where to begin?

I want to make container gardening simple and fun for everyone, but I need your help to reach more people.

Most readers decide to get a book based on what others say. If you found this book helpful, a quick review could...

...help one more person discover the joy of growing their herbs.
...help one more family share healthy homegrown veggies.
...help one more balcony turn into a mini garden.
...help one more beginner feel confident with their plants.
...help one more community grow greener and healthier.

If you love gardening and helping others get started, thank you for being a part of this journey with me!

- **Kiley Manning**

Chapter 5

Gardening Technology and Tools

In today's app-driven world, gardening apps are essential tools for novices and experienced gardeners. These apps and online social circles provide valuable insights into plant care, pest diagnosis, and more.

5.1 Digital Space for Gardening

Different digital apps can help you plan and manage your garden. You can stay informed and learn about different plants and even techniques that can give you a boost in caring for and maintaining your garden.

Gardening Apps

Imagine having a plant identifier in your pocket, where a simple photo snap gives you the name and complete care guide for any plant. Some gardening apps use AI to offer this precise service, making plant identification both fun and educational.

Beyond identification, some gardening manager apps can also serve as digital caretakers, reminding you when it’s time to water, fertilize, or repot your plants. These reminders are particularly useful if you have a large collection of containers, each with its own care schedule.

For pest and disease management, apps like Plant Identifier provide instant access to diagnostic tools and solutions, helping you address issues before they escalate.

Making use of these apps for your gardening routine can help boost plant health and improve your gardening results, such as bigger or abundant blooms and better harvests.

Virtual Garden Planning Tools

When it comes to planning a garden, the amount of information available can get overwhelming, especially when you consider that you still need to lay out your container garden space.

Tools like garden planners allow you to design and visualize your garden in a digital space, experimenting with different layouts, plant combinations, and even hardscaping options before committing to physical work.

These tools are handy for container gardeners as they allow you to maximize limited space by organizing your pots and planters visually to ensure each plant receives adequate light and space. These often include features like crop rotation alerts and seasonal change simulations, helping you plan for the immediate season and year-round beauty and productivity. That way, you won't lose precious time planning your container garden while making the most of every element of your garden, from a more efficient and beautiful layout to methodical seasonal transitions.

Online Communities

No gardener is an island, and online gardening communities are invaluable for sharing knowledge, seeking advice, and finding inspiration.

Below are some types of online communities you can utilize to grow your knowledge:

- **Online Forums and Groups:** Getting into online communities with designated threads and forums, like the Gardening subreddit, connects you with fellow gardening enthusiasts, offering a space to exchange tips, celebrate successes, and seek solutions to challenges.
- **Gardening-Specific Platforms:** These platforms will allow you to engage with local gardeners, which is incredibly helpful for understanding regional planting tips and weather challenges. Socializing and participating in these gardening circles can also lead to discovering new gardening trends or rare plant varieties, which could help improve and broaden your garden's diversity and appeal.

The collective knowledge and wisdom shared in these forums often lead to innovative solutions you might not find in traditional gardening books, making them a dynamic resource for continuous learning and improvement.

5.2 Modern Technological Advancements

Other than digital advancements that can help keep you informed and updated about caring for your container garden, modern gardening appliances and accessories can assist you in keeping track of your individual plants' health and growth and streamline your gardening experiences.

Smart Sensory Gardening Systems

Including smart sensors and automated systems in your garden can significantly reduce the guesswork involved in plant care. Devices like soil moisture sensors, light meters, and innovative irrigation systems can provide real-time data and automate watering schedules for optimal growing conditions.

- **Smart Pots:** These are equipped with sensors to monitor soil moisture, temperature, and nutrient levels and can send alerts to your smartphone, allowing you to provide precise care for each plant and ensure they thrive even in the harshest conditions.

- **Robotic Gardeners:** These can take on some of the more tedious gardening tasks, like weeding and maintaining garden beds autonomously, allowing you to focus on planning and enjoying your garden.

- **Weather Monitoring Tools and Apps:** These can provide hyper-local weather forecasts and alerts, helping protect your plants from adverse weather conditions.

By staying informed about temperature changes, rainfall, and potential storms, you can take proactive measures to safeguard your garden.

Self-Watering Containers

Micromanaging your garden can be troublesome when you forget to water your plants, so self-watering containers have slowly become a staple in most container gardens. These offer a low-maintenance solution for keeping plants hydrated while preventing root rot.

These containers feature a shallow water reservoir, allowing the soil above to draw water steadily through capillary action or wicking systems.

The consistent moisture availability of self-watering containers is particularly beneficial during hot weather or for busy gardeners, reducing manual watering frequency.

Self-watering containers come in various designs, from commercially available pots to customizable systems for different plant sizes and types.

With this in mind, here are some of the best practices in setting up self-watering containers to maintain the quality and health of plants:

- **Potting Mix**: Use a well-aerating mix with good wicking properties. Peat moss or coconut coir mixes are recommended for maintaining moisture without becoming waterlogged.
- **Water Level**: Regularly check the reservoir and refill it before it runs dry to maintain the capillary action. Consistent water levels help ensure plants receive a steady supply of moisture.
- **Cleaning**: Clean the reservoir periodically to prevent algae and bacteria buildup, which can hinder the system's efficiency and harm plant health.
- **Avoid Overfilling**: Overfilling the reservoir can lead to water-saturated soil, increasing the risk of root rot, especially for plants sensitive to excessive moisture.
- **Top Watering**: Occasionally water from the top to flush out any salt accumulation from fertilization. This helps maintain soil health and nutrient balance.

Using self-watering containers offers several benefits that enhance both the gardening experience and plant health:

- **Reduced Maintenance**: Less frequent watering means less daily attention is required, making it easier to maintain a garden amidst a busy schedule.
- **Water Efficiency**: These containers optimize water use, reducing wastage and ensuring that plants receive only what they need.
- **Consistent Growth**: Steady moisture availability promotes healthier, more robust plant growth by minimizing the stress caused by irregular watering.
- **Convenience**: Ideal for urban gardening and spaces where traditional watering methods might be inefficient.

To fully reap the benefits of these smart containers, here are some considerations for optimal use.

- **Climate Adaptation**: In extremely hot climates, additional shading or cooling measures might be necessary to prevent excessive water evaporation.
- **Seasonal Adjustments**: Adjust the use of self-watering containers seasonally, particularly during periods of heavy rainfall or drought, to maintain optimal plant health.
- **System Customization**: Tailor self-watering systems to the specific needs of your plants, considering factors like root depth and water consumption rates.

As wonderful as self-watering containers sound, it's highly important to note that *not all* containers are suitable for all plants.

Self-watering containers are not exempt from problems, and most of them come from the benefit of watering from the *bottom* of the container rather than from the top.

Here are some common problems that come with self-watering containers, especially when misused and without proper maintenance:

- **Sensitivity to Overwatering**: Some plants are sensitive to consistently high moisture levels and can easily suffer from root rot or fungal diseases. These plants might prefer a drier environment than traditional watering can provide.
- **Waterlogging**: Self-watering containers often have a reservoir that can sometimes lead to waterlogging if not appropriately managed. Plants that require well-drained soil may not thrive in such conditions.
- **Root System Differences**: Plants with deep, extensive root systems may not benefit from the shallow watering provided by self-watering containers. They might need deeper soil to spread their roots effectively.
- **Nutrient Distribution**: In self-watering containers, nutrients might not distribute evenly, which could be problematic for plants that rely on consistent nutrient uptake from the soil.
- **Growth Habit**: Certain plants, especially those that grow rapidly or have spreading habits, might outgrow the confines of a self-watering container quickly, leading to root-bound conditions.

Given the advantages and disadvantages of self-watering containers, only certain plants can survive much better in self-watering containers due to their preference for consistent moisture and suitability with bottom-type watering.

Such plants that thrive with self-watering containers are:

- **Tomatoes**: These fruiting plants benefit greatly from steady moisture levels, which help prevent blossom end rot and other water-stress-related issues.
- **Peppers**: Like tomatoes, peppers require regular moisture to produce healthy fruits and avoid stress.
- **Herbs**: Herbs such as basil, mint, and parsley thrive in consistently moist conditions, making them ideal for self-watering systems.
- **Tropical Houseplants**: Plants like ferns, peace lilies, and other tropical varieties prefer a constant moisture supply, which self-watering containers can provide.

LED Grow Lights

LED grow lights are a game-changer for indoor gardening. They can turn dimly lit spaces into fertile grounds for lush, vibrant plant growth by copying the spectrum of sunlight essential for emulating photosynthesis.

Using LED grow lights can extend the growing season and maintain plant health, even in places with low or limited natural light.

Given that these types of lights would require a good bit of electricity, energy-efficient LED grow lights are available and are known to consume less electricity; however, these produce less heat compared to traditional grow lights like HID or fluorescent bulbs.

Still, these efficient lights reduce both your utility bills and the environmental impact of your gardening activities.

LED grow lights don't produce much heat, so you can place these lights closer to plant canopies without the risk of burning them, increasing light intensity and effectiveness. This proximity encourages stronger, healthier growth and faster maturation rates, which is excellent for fruiting and flowering plants that require high light levels.

To fully understand the importance of these LED grow lights and how they interact with different plants in your garden, here's a brief review of how photosynthesis works.

Photosynthesis utilizes water and carbon dioxide to produce glucose. It emits life-sustaining oxygen—a process which occurs in the chloroplasts within plant cells and is essential for their growth and energy production.

- **Light Absorption:** Chlorophyll in chloroplasts absorbs light, primarily in the blue and red wavelengths.
- **Water Splitting:** Light energy splits water molecules into oxygen, protons, and electrons.

- **Energy Conversion:** The energy from light is converted into different forms of chemical energy: ATP and NADPH.
- **Sugar Production:** ATP and NADPH convert carbon dioxide into glucose during the Calvin cycle.

LED grow lights are, therefore, the perfect companion, as they can supply specific light spectra needed for photosynthesis so that plants can receive the optimal light for their growth stages.

Selecting the right LED grow lights depends on several factors, including the types of plants, their growth stage, and your indoor setup specifics.

Consider the following:

- **Light Spectrum:** Plant growth requires different spectra for different stages.
    - **Blue Light (400-500 nm):** Promotes vegetative growth. Ideal for leafy greens and seedlings.
    - **Red Light (600-700 nm):** Benefits the flowering stage. Crucial for fruiting plants like tomatoes and peppers.
    - **Full Spectrum:** Adjustable or full-spectrum LEDs cater to all growth stages.
- **Light Intensity**:
    - **High-Light Plants:** Tomatoes, peppers, and cannabis require more intense light.
    - **Low-Light Plants:** Ferns and some herbs require less intense light.
- **Coverage Area**: Ensure the number of lights matches the size of your grow area to provide uniform light distribution.

Once you've chosen your LED grow lights and set them up, pay special attention to the distance from the lights to your plants and the colors of the materials your plants are surrounded by. Make sure these grow lights are on a timer.

Generally, lights should be placed closer to the seedlings, about 2-4 inches (5-10 cm) away, and raised as plants grow. Keeping the light distribution even is crucial, especially in larger setups. Reflective materials like Mylar or white paint can help maximize light efficiency.

Lastly, maintain a consistent light schedule that mimics natural daylight patterns, usually between 8-10 hours of light per day (some plants need 12-16 hours, while fruiting plants would need 18 hours).

Timers can help simulate light cycles, providing consistent and predictable light exposure essential for healthy plant growth.

Combining Natural and Artificial Light

If possible, it's best to combine natural and artificial light to maximize growth and energy efficiency. Supplement natural light with LED grow lights to ensure all plants receive optimal light.

This hybrid approach balances natural resources with technology, maintaining consistent growing conditions, which is particularly beneficial during shorter daylight hours.

A common problem with LED grow lights is that your plants can either get too much or too little light. Most plants require a good amount of light, but having *too* much of it can give your plants something similar to a sunburn, which can be harmful and potentially kill your plant if it's not noticed earlier.

Here are some easy-to-spot warning signs to start optimizing the amount of light your plants are receiving:

- **Too Much Light:**
    - **Leaf Burn:** Yellow or brown patches on leaves.
    - **Bleached Leaves:** Leaves may turn white, indicating light bleaching.
    - **Wilting:** Despite adequate water, plants may wilt due to excessive light stress.
- **Too Little Light:**
    - **Leggy Growth:** Stems grow long and thin as plants stretch toward the light.
    - **Pale Leaves:** Leaves may turn pale or yellow due to insufficient chlorophyll production.
    - **Slow Growth:** Overall, plant growth may be stunted and slow.
- **Optimal Light:**
    - **Healthy Leaves:** Deep green, robust leaves.
    - **Compact Growth:** Sturdy stems and compact, bushy growth.
    - **Vibrant Colors:** Flowers and fruits display vibrant, rich colors.

And finally, here are some additional tips to effectively use and maintain your LED grow lights and plants.

1. **Monitor Plant Response:** Regularly check your plants for signs of too much or too little light, such as leaf burn or excessive stretching, and adjust the light distance accordingly.

2. **Maintain Clean Lights:** Keep the surface of your LED grow lights clean to ensure maximum light penetration. Dust and grime can reduce light efficiency.
3. **Consider Light Spectrum Adjustments:** Use lights with adjustable spectra to cater to different growth stages and optimize plant development.
4. **Temperature Management:** While LEDs emit less heat, ensure your grow room is well-ventilated to avoid heat build-up, especially in enclosed spaces.

You gain control over your garden's environment by utilizing LED grow lights, allowing you to nurture a diverse range of plants year-round.

When combining natural and artificial light, it's also crucial to pay close attention to the heat zones of the area you're living in if you want to give your plants enough natural sun while avoiding potential threats like overheating, bleaching, sunburn, and other heat-related problems that you and your container gardening plants could encounter. Summer can also pose a possible heat risk if this is not considered.

Hydroponics and Aquaponics

Hydroponics and aquaponics are innovative gardening methods that allow efficient, soil-free plant cultivation.

Both techniques offer unique advantages, including water conservation, faster plant growth, and the potential for more abundant yields compared to traditional gardening methods.

The main difference between the two water gardening methods is that one uses nutrient-rich water, while the other involves fish to help plants grow.

Getting Started with Hydroponics

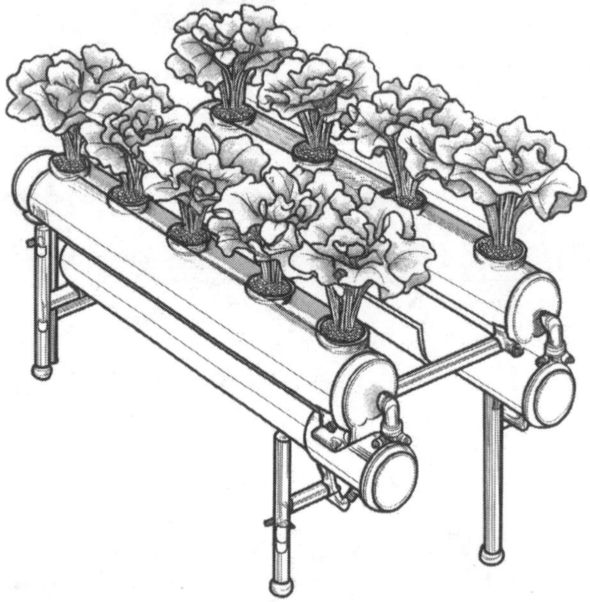

This method doesn't require soil to grow plants. Instead, it uses a nutrient-rich water solution to provide all necessary sustenance for plant growth.

This technique conserves water and space, leading to faster growth and potentially higher yields. For beginners, we highly suggest starting with simple systems like the wick system or deep-water culture:

- **The Wick System:** Uses a wick to draw nutrient solution from a reservoir to the plant's root zone. It is perfect for smaller, less water-hungry plants like herbs. This setup is easy to put up and requires very little maintenance.
    - **Ideal Wick System Plants:** Basil, mint, and lettuce.

- **Deep Water Culture Systems:** Plants are suspended in a solution that allows their roots to be directly immersed in a nutrient mix. It is well-suited for larger, faster-growing plants. DWC systems can be easily constructed using typical containers modified to hold nutrient solutions and support plant growth.
    - **Ideal DWC plants:** Tomatoes, peppers, and spinach
- **Nutrient Film Technique**: Plants thrive within a channel where a thin film of nutrient solution flows over their roots continuously. This method ensures that roots have consistent access to nutrients, oxygen, and water.
    - **Ideal NFT plants:** Strawberries, arugula, and kale
- **The 'Ebb and Flow' Method:** Also known as the Flood and Drain, this method involves periodically flooding the grow bed with nutrient solution and draining it back into the reservoir. This cycle ensures that plants receive adequate nutrients while allowing roots to aerate.
    - **Ideal Ebb and Flow Plants:** Cucumbers, bell peppers, and eggplants

Hydroponics can seem a bit confusing to set up at the start, but the routine maintenance can be pretty simple once you get the hang of it.

As long as you keep maintenance consistent, you're on the track to growing happy, healthy plants without having to worry too much about plant threats and diseases.

Introduction to Aquaponics

Aquaponics combines plant cultivation with aquaculture, where fish waste provides nutrients for plants, purifying the water that returns to the fish. Fish species such as tilapia, goldfish, and koi are popular choices due to their ability to survive and thrive in various conditions.

Setting up a basic home aquaponic system involves integrating a fish tank and a plant-growing bed in a recirculating environment.

Here are some aquaponics systems that you can apply to your container garden:

- **Media Bed System:** A simple system that involves growing plants in a bed of gravel or clay pellets, which also act as a biofilter. The media bed system is an easy entry point into aquaponics, requiring minimal setup and maintenance.

  - **Ideal MBS Plants:** Leafy greens and herbs (e.g., lettuce, basil

- **Deep Water Culture System** or **Raft System**: This system may sound like the hydroponics DWC system, but the main difference is that it has fish. Plants are placed on floating rafts with their roots submerged in nutrient-rich water, making this system more appropriate and efficient for large-scale production.
  - **Ideal Raft System Plants**: Leafy greens (e.g., kale, lettuce)
- **Nutrient Film Technique**: Just like the hydroponic NFT system, this method involves a thin film of water, this time enriched with nutrients from fish waste flowing over the plant roots.
  - **Ideal NFT Plants:** Herbs (e.g., parsley, cilantro).

The key to successful aquaponics lies in maintaining the right fish-to-plant ratio for a balanced and sustainable system.

Maintaining and Monitoring Systems

Like with any gardening system, regular maintenance and monitoring of the water pH, nutrient levels, and system health are crucial for the health and productivity of both hydroponic and aquaponic systems:

- **Hydroponic Maintenance:**
  - **Water pH:** Use a pH meter or test kit to check if the nutrient solution is slightly acidic, typically 5.5 to 6.5 pH for hydroponic systems.
  - **Nutrient Levels:** Frequently replenish and adjust nutrient solutions to match plant growth stages. Use conductivity meters to measure nutrient concentration in hydroponics

- **Aquaponics Maintenance:**
    - **Water pH:** Maintain a pH between 6.8 and 7.0 to accommodate both plants and fish.
    - **Nutrient Levels:** Monitor your fish's health and feeding rate to maintain nutrient balance, observe fish behavior, and adjust feedings accordingly to maintain water quality.

Both hydroponic and aquaponic maintenance require system health check-ups using tools like pH and conductivity meters. These meters help fine-tune the environment for optimal plant and fish health.

Install sensors to continuously monitor water parameters so you can be promptly alerted to significant changes, which can help you consistently oversee and identify potential issues earlier on.

Chapter 6

Vegetables and Herbs

In the previous chapters, we explored the essential foundations, design strategies, and container gardening techniques. Understanding these principles is crucial for making informed decisions about what to grow and setting up a thriving garden.

With this knowledge, you can confidently plan and create your ideal container garden. So for this chapter, we'll dive into the exciting process of growing a variety of plants, starting with vegetables and herbs.

These versatile and rewarding choices not only enhance your gardening skills but also provide fresh, organic ingredients for your kitchen.

6.1 Growing Salad Greens

Growing salad greens in containers is a rewarding way to enjoy fresh, homegrown leaves all year round. Whether you prefer crisp lettuce, tender spinach, or nutrient-rich microgreens, container gardening offers the flexibility to cultivate these greens with minimal effort.

You can ensure a steady supply of nutritious, flavorful green salads for your kitchen with the proper techniques.

Salad Green Varieties

Selecting the right varieties for the season is crucial to ensure a successful harvest:

- **Warm Months:** Choose heat-tolerant varieties like oak leaf, butterhead, or romaine lettuce. These greens are less likely to bolt (go to seed) during high temperatures, ensuring longer-lasting crops.
- **Cool Months:** Cold-hardy greens such as spinach, kale, and arugula thrive in cooler temperatures. They can withstand light frost with minimal protection, making them ideal for autumn and early spring planting.
- **Year-Round Options:** Microgreens and baby greens can be grown year-round, indoors or outdoors. Their quick growth cycle makes them a perfect choice for continuous harvests.

Once you have chosen your preferred variety of salad greens, here are some tips to ensure that they grow efficiently and to have a continuous supply of them:

- **Seasonal Adjustments:** Understanding the seasonal preferences of your salad greens is essential for optimizing growth. Provide shade for heat-sensitive varieties during summer and protect frost-tolerant plants in winter with simple coverings or by moving containers to sheltered locations.
- **Rolling Harvest:** For a continuous supply of fresh greens, employ a rolling harvest strategy by planting seeds every two to three weeks. This staggered planting ensures that new crops are always maturing.

Container and Soil Preparation for Salad Greens

Use containers at least 6–8 inches deep with multiple drainage holes to prevent waterlogging. Materials like plastic, terracotta, or fabric pots are all suitable, with fabric pots offering improved aeration and drainage. Refer to Chapter 1.1: *Proper Drainage* for guidance on adding drainage holes to containers.

Meanwhile, a well-draining potting mix is crucial for healthy salad greens. A blend of peat moss, perlite, and vermiculite ensures proper hydration and prevents water saturation. Incorporating compost or organic matter adds essential nutrients for vigorous growth.

Caring for and Maintaining Salad Greens

Salad greens are relatively easy to care for because of their fast growth, minimal needs, and adaptability. Still, it's important to take care of them just as you would any other garden plant:

- **Watering Routines:** Keep the soil consistently moist but not soggy. Self-watering containers can help maintain optimal moisture levels, reducing the risk of under- or over-watering.
- **Light Requirements:** Salad greens thrive with 6–8 hours of sunlight daily. For indoor gardening, supplement natural light with grow lights to ensure robust growth.
- **Fertilizer Application:** Feed salad greens regularly with a nitrogen-rich, water-soluble organic fertilizer every two weeks. Alternatively, a slow-release fertilizer provides steady nutrition throughout the growing season.

- **Pest Management:** Inspect plants regularly for pests like aphids or slugs. Address infestations promptly with organic methods such as insecticidal soap or neem oil.
- **Companion Planting:** Herbs like chives and garlic near your salad greens can deter pests, while flowers like marigolds attract beneficial insects. This natural approach enhances the health and productivity of your garden.
- **Harvesting Techniques:** Depending on the type of salad greens you're harvesting, there are two methods you can apply:
    - **Cut-and-Come-Again Method:** Harvest the outer leaves about an inch above the base, allowing the inner leaves to grow. This technique is ideal for lettuce, spinach, and kale.
    - **Microgreens Harvesting:** Microgreens are typically harvested once by cutting the plants above the soil line when they reach 2–3 inches tall. Use sharp scissors or knives to make clean cuts and minimize plant stress.

Growing salad greens in containers is both practical and rewarding. By selecting the right varieties, using proper planting techniques, and maintaining consistent care, you can enjoy a continuous supply of fresh greens that elevate your meals and promote a sustainable lifestyle.

6.2 Essential Kitchen Herbs

Growing herbs in containers not only enhances your home's beauty and fragrance but also provides the convenience of having fresh ingredients like basil, cilantro, and mint right at your fingertips. A kitchen herb garden is a delightful addition that combines functionality with aesthetics, making it easy to elevate your culinary creations while enjoying the therapeutic benefits of gardening.

Choosing Herbs

Select herbs that are easy to grow, versatile in the kitchen, and suited for container gardening:

- **Basil**: Basil is a must-have in any herb garden with its lush, aromatic leaves. It thrives in warm conditions with plenty of sunlight. Varieties like 'Genovese' basil are particularly well-suited for containers due to their compact growth.
- **Cilantro**: Favored for its fresh, citrusy flavor, cilantro grows best in cooler temperatures and is ideal for spring and fall planting. 'Slow Bolt' cilantro is great for extending its growing season.
- **Mint**: A robust grower, mint should be planted alone as it tends to spread aggressively. Compact varieties like *spearmint* or *peppermint* are perfect for container gardening. Its refreshing flavor makes it a versatile addition to drinks, desserts, and savory dishes.

These herbs enhance your recipes and add vibrant greenery to your kitchen space, doubling as decorative plants.

Container and Soil Preparation for Herbs

Choose containers with adequate drainage holes to prevent waterlogging. Traditional clay pots are excellent for their porous nature, but plastic or glazed pots work well if they allow proper drainage.

The container size should match the root system of the herb—for example, deeper containers for basil and wider ones for mint.

As for the soil mix, a high-quality potting mix with peat moss, perlite, and vermiculite ensures good drainage and aeration, preventing root rot.

Adding compost or organic fertilizer enriches the soil with nutrients, promoting healthy growth.

Caring for and Maintaining Herbs

Much like salad greens, herbs are also reasonably simple to care for and maintain, especially since they are versatile, require minimal space, and are adaptable.

- **Watering Routines:** Each herb has specific watering needs:
    - **Basil and Cilantro:** Require consistently moist soil. Water deeply, but let the topsoil dry slightly between waterings to avoid overwatering.
    - **Mint:** Tolerates some dryness but benefits from regular watering to maintain lush growth.
- **Light Requirements:** Most herbs need 6–8 hours of sunlight daily:
    - **Outdoors:** Place containers in areas with morning sun and afternoon shade to prevent heat stress.
    - **Indoors:** Position containers on a sunny windowsill or use grow lights to ensure adequate light.
- **Temperature and Humidity**: Different herbs require different temperatures and humidity for optimal growth:
    - **Basil and Cilantro:** Thrive in temperatures of 65–75°F (18–24°C) but may become stunted if temperatures drop below 50°F (10°C).

    - **Mint:** Prefers similar temperatures but can tolerate cooler conditions. Ensure roots are protected from frost, as leaves and stems may die back below 32°F (0°C).

- **Pruning:** Regular pruning encourages bushier growth and prevents flowering, which diverts energy from leaf production.
    - **Basil and Cilantro:** Pinch off the tips of basil and cilantro to promote branching
    - **Mint Leaves:** Harvest mint leaves frequently to maintain their flavor.
- **Pest Management:** Check for common pests like aphids, slugs, or spider mites. Use organic solutions like neem oil or insecticidal soap. Companion planting herbs such as chives and flowers like marigolds can deter pests while attracting beneficial insects. Exercise caution with methods like diatomaceous earth, ensuring it doesn't harm pollinators.

Using and Preserving Fresh Herbs

Fresh herbs add vibrant flavors to your cooking. Depending on what you're preparing, here are some common herbs along with the respective dishes you can add them to:

- **Basil:** Tear fresh leaves over pizzas, pasta, or Caprese salads.
- **Cilantro:** Chop it into salsa, curries, and marinades for freshness.
- **Mint:** Add to beverages, desserts, or savory dishes for a refreshing twist.

Unlike salad greens, herbs such as basil, cilantro, and mint can be preserved to extend their shelf life. Many of these herbs can be preserved in different ways.

Preservation methods for herbs include:

- **Drying:** There are three different drying methods you can try to keep your herbs for a longer time.
    - **Air Drying**: Tie herb bunches and hang them upside down in a warm, airy space until the leaves crumble easily. Store in airtight containers to preserve their flavor.
    - **Oven Drying**: Spread herb leaves on a baking sheet and place them in a low-temperature oven until completely dry. Store in airtight containers.
    - **Dehydrator**: Use a food dehydrator to dry the herbs evenly and efficiently.
- **Freezing:** When freezing herbs, there are two methods you can go for.
    - **Herb Cubes:** Chop herbs, pack them into ice cube trays with water or olive oil, and freeze. Transfer the cubes to freezer bags for easy use in soups, sauces, or stews.
    - **Individual Leaf Freezing**: Rinse and dry the herbs, then spread them in a single layer on a baking sheet. Freeze until solid, then transfer to freezer-safe bags or containers.

Growing herbs in containers is a simple and rewarding way to bring freshness and flavor into your home. With proper care and attention, your kitchen herb garden can thrive year-round to provide endless additions to your food recipes.

6.3 High-Yield Vegetables

Growing turnips, tomatoes, peppers, and cucumbers in containers maximize your gardening space and brings the delight of harvesting fresh vegetables right from your balcony or patio.

These plants are known for their bountiful production, but achieving such yields in a confined space requires careful consideration of the containers you use.

Container and Structure for High-Yield Vegetables

Selecting suitable containers is crucial to accommodate the robust root systems of these vegetables, ensuring they have ample room to grow without becoming root-bound.

Here are some guidelines:

- **Cucumbers and Peppers**: Use containers that are at least 5 gallons (18.9L) in size.
- **Tomatoes**: Preferably grown in 10-gallon (37.85L) or larger containers, depending on the variety.
- **Material**: Fabric pots offer excellent aeration and prevent overheating, which are crucial for root health. Plastic or ceramic pots can also be suitable if they have sufficient drainage holes to avoid waterlogged soil.

Unlike herbs and vegetables, high-yield vegetables would need support structures to climb on.

Use stakes, cages, or trellises to support the growth of these vegetables, especially tomatoes and cucumbers, which tend to sprawl on the ground. This also helps improve airflow and reduces the risk of disease and infestation.

Caring for and Maintaining High-Yield Vegetables

High-yielding vegetables can be bountiful and are usually considered "low-maintenance." Despite these vegetables' resilience and strong adaptability, it's still advisable to take care of them like you would any other.

- **Watering Routines**: Proper watering and feeding of high-yield vegetables are pivotal for maximizing production:
    - **Tomatoes and Cucumbers:** These require consistent moisture. Apply a mulch layer on the soil surface to help retain water and allow the soil to dry slightly between watering to encourage healthy root growth. Overwatering can lead to root diseases.
    - **Peppers:** More drought-tolerant but still need regular watering for plump, healthy fruits.
- **Light Conditions**: Like the previous greens, have your containers placed in spots where your plants can get at least 6-8 hours of full sunlight every day, especially since these vegetables thrive in bright conditions.
- **Fertilization:** Use a balanced, slow-release fertilizer to supply nutrients steadily. Adding supplemental liquid fertilizer high in potassium and phosphorus during the flowering stage can help boost fruit production.

- **Pest Management**: High-yielding vegetables such as tomatoes, cucumbers, and peppers they need effective pest and disease management to remain healthy and productive as they grow:
    - **Tomatoes:** Susceptible to blights and fungal infections. Make sure there's good airflow and avoid watering above this plant—water directly into its soil.
    - **Cucumbers:** Can become infested with cucumber beetles and suffer from powdery mildew. Floating row covers can protect cucumber plants from pests early in the season, and applying a milk solution and water can prevent mildew.
    - **Peppers:** Could face issues like blossom end rot, which is often linked to calcium deficiency and irregular watering. Maintain a consistent watering schedule and use calcium supplements to prevent this condition.

Rotation and Companion Planting

Rotating your crops each season will prevent soil depletion and plant diseases. Companion planting can also help deter pests and improve plant health.

For example, basil planted near tomatoes can repel certain pests and enhance tomato flavor. You can check Chapter 2.3: *Companion Planting*, under the *Herb and Vegetable Companions* section for more combinations.

6.4 Root Vegetables

Growing root vegetables such as carrots, beets, and radishes in containers offers a unique and rewarding gardening experience. Even with limited space, you can enjoy the crisp and fresh flavors of homegrown produce.

The successful cultivation of these vegetables hinges on selecting the appropriate container depth.

Many gardeners assume that all root crops are high-yielding, but this isn't always the case. The key difference lies in their growth rate and harvest potential. Certain root vegetables, like potatoes and onions, are known for being highly productive because they grow quickly and produce abundant harvests in a short time.

On the other hand, root crops like parsnips, wild celery, and horseradish take longer to mature and typically yield smaller harvests. Understanding these variations allows you to plan your garden more effectively, combining fast-growing crops with slower-yielding varieties for a balanced and rewarding harvest.

Containers and Soil Preparation for Root Vegetables

Root vegetables need ample space to expand and grow downwards. Each root crop will have different requirements in terms of depth, so here are some guidelines:

- **Carrots**: A container depth of at least 12 inches is ideal.
- **Beets and Radishes**: These would require slightly less depth than carrots but should still be planted in containers at least 8-10 inches (20.32-25.4 cm) deep.

Choosing the right container size impacts the health and yield of these vegetables. Placing them in shallower containers will stunt their growth and lower the quality of their flavor and texture.

As for soil composition, the quality would determine the success of your harvest. Root vegetables thrive in loose, well-draining soil for easy root penetration. Here's how to prepare soil that's perfect for root crops:

- **Soil Mix**: Use a blend of garden soil, compost, and a bulking agent like perlite or sand.
- **Adding Compost**: Enriches the soil with nutrients and helps keep it loose and fertile.
- **Moisture Level**: Regularly check the soil's moisture level; it should be moist but not soggy. Ensure water drains freely from the bottom of the container.

Planting and Thinning

Root vegetable seeds must be sown directly into their final containers as they don't respond well to transplanting.

Follow these steps for the right way to plant and thin root crops:

- **Sowing Seeds**: Distribute seeds evenly across the soil surface, then cover with a thin layer of finely sifted soil.
- **Thinning Seedlings**: As seedlings begin to sprout, thinning is necessary to prevent overcrowding. To thin seedlings, you need to grasp the sprout near the root and pull them up.
    - For carrots and radishes, space seedlings about 2-3 inches (5-7 cm) apart.
    - Beets may require slightly more space, about 3 to 4 inches (7-10cm) between plants. After thinning, it's best to hill up the soil around the sapling for extra support.

Thinning is essential for growing healthy and well-formed root vegetables. As for the thinned sprouts, you can either use them as compost or use them as salads or garnishes.

Harvesting

Each root vegetable usually needs to be harvested by determining the diameter of the top, and root vegetables that are harvested at the right time mean better quality with the best flavor and texture. Here are some harvesting tips:

- **Carrots**: When their tops are about 3/4 inch (1.95cm) in diameter, they're usually ready for harvest.
- **Beets**: Best harvested when they reach about 2 to 3 inches (5-7cm) in diameter.
- **Radishes**: These root crops mature very quickly and should be harvested as soon as they reach edible size, usually within a month of planting.

To harvest these root vegetables, gently loosen the soil around them with a fork or hand trowel, then pull the root vegetable out by the base of its stems. Handling the vegetables gently reduces damage and prolongs their shelf life.

Additional Vegetable and Herb Growing Tips:

- **Successive Planting**: Planting new seeds every few weeks to secure a continuous harvest throughout the growing season.
- **Companion Planting**: Pair root vegetables with companion plants that can help deter pests and improve growth, such as planting onions or garlic near carrots to repel carrot flies
- **Monitoring**: Keep an eye on your vegetables for any signs of pests or diseases, and make sure to address these issues immediately to ensure a healthy crop.

Chapter 7

Fruits and Berries

One of the best things you could be greeted with would be the sweet fragrance of ripe strawberries cascading beautifully from their hanging baskets. This picturesque scene isn't just from a gardening magazine; it's a snapshot of what you can cultivate in your urban space, no matter the size.

7.1 Strawberries in Baskets

Growing strawberries in hanging baskets is a delightful way to produce fresh, succulent fruits and a practical solution for limited space, adding aesthetic value to your balcony, patio, or even a sunny window.

Picking Strawberry Varieties

Selecting the right strawberry varieties is crucial for success and yield. If you're growing strawberries indoors, then everbearing and day-neutral strawberries are ideal for hanging baskets.

These berries produce fruit throughout the growing season, unlike June-bearing varieties that fruit once a year.

There are several varieties of everbearing and day-neutral strawberries. The ones listed here are relatively easier to take care of, especially for novice gardeners:

- **Everbearing Varieties:** The varieties listed below provide a steady supply of berries from early summer until fall. These two varieties love a full sun.
  - **Quinault Strawberry:** Produces an abundance of large, sweet strawberries. The Quinault's ripe berries can be gathered from late spring to mid-fall.
  - **Ozark Beauty:** Known as the best variety among everbearing ones, it consistently produces fruit in the spring and late summer or fall.
- **Day-Neutral Varieties:** These plants fruit throughout their respective growing seasons as long as temperatures remain between 35°F (1.6°C) and 85°F (29°C), maximizing vertical space and providing fresh berries continuously. These also need full sun.
  - **Albion:** Grows and produces crops from hot summers until late fall and is one of the easiest day-neutral varieties to cultivate and care for.
  - **Seascape:** Produces ripe berries from late May to September. However, this strawberry variety would need to be watered more frequently.

These varieties are perfect for hanging baskets as their continuous yield offers fresh berries for your table or a beautiful, edible addition to your home decor.

Soil Mix and Basket Setup

Use a high-quality potting mix enriched with compost for essential nutrients. When planting strawberries, position their crowns just above the soil surface to prevent decay and space them evenly to allow adequate air circulation and healthy growth.

Choose baskets that are at least 12 inches (30.48 cm) in diameter to make sure there's enough room for roots and plant growth, and hang the baskets in locations that receive at least 6-8 hours of sunlight daily for the best fruit yield.

Setting up hanging baskets for strawberries requires attention to detail. Make sure your baskets have proper drainage to prevent waterlogging. Line the baskets with coir or sphagnum moss to retain moisture while allowing surplus water to drain.

Caring for and Maintaining Strawberries

Strawberries are generally a little more difficult to take care of—and that's particularly true for the June-bearing variety, which is commonly sold in markets. However, the variety we'll focus on discussing proper care for is the everbearing and day-neutral strawberries, as these are easier to maintain and grow.

- **Watering Routines**: Proper watering is crucial, as hanging baskets tend to dry out faster than ground soil, particularly in warm weather. Make sure to water your strawberries daily or even twice daily during hot spells, ensuring the soil is moist but not soggy.
- **Fertilization and Mulching**: Feed your strawberries every other week with a balanced liquid fertilizer, which can help promote vigorous growth and bountiful fruiting, and layer mulch on top of the soil to retain moisture and stop weeds from spreading.
- **Pruning**: Remember to regularly check for and remove any dead or diseased leaves to keep the strawberry plants healthy and prevent fungal infections. Also, prune the runners—long stems that grow out from the main plant—to encourage more fruit production, as the plant's energy is directed toward creating strawberries rather than spreading.
- **Pest Management**: It's best to use organic pest control methods and techniques that can protect them from common garden pests. You can also try covering your strawberries with lightweight fabric to stop insects from laying eggs on them.

Harvesting and Renewal

Harvest your strawberries when they are fully red and the seeds on the outside are dark to make sure they're of the best flavor and quality.

Instead of pulling on the fruit, gently twist or cut the stem above the berry to avoid damage.

Strategically harvesting these plants regularly will encourage them to produce more fruit, extending the yield period.

At the end of the season, it's good practice to renew your plants by replacing the soil in the baskets and adding new strawberry plants every few years to prevent disease build-up and keep your baskets productive and vibrant season after season.

Store newly harvested strawberries in a cool place, and consume them within a few days for the best taste. Consider rotating your strawberry plants every 2-3 years to maintain soil health and reduce the risk of diseases.

7.2 Dwarf Fruit Trees

Growing large fruit trees would be impossible in container gardening, much less in urban spaces with limited to no space. That's why dwarf fruit trees are perfect for these limited spaces and container gardening.

The dwarf trees' small size makes them easier to manage; pruning, spraying, and harvesting can be done standing on the ground without the need for ladders. They can reach maturity faster than regular fruit trees and produce fruit within one to three years of planting.

This accessibility is especially appealing for those with limited mobility or beginners who might find managing large trees daunting.

Choosing Dwarf Tree Varieties

There are several dwarf fruit trees to choose from. The ones listed below are some of the easier ones to care for:

- **Dwarf Apple Trees**: Varieties like 'Cameron Select' and 'Stayman Winesap' are compact and produce delicious apples.
- **Dwarf Lemon Trees**: The Dwarf Meyer Lemon Tree is a popular choice, offering juicy, slightly sweet lemons.
- **Dwarf Cherry Trees**: The 'Stella' cherry tree is self-pollinating and produces dark red, sweet cherries

Container and Soil Preparation

When selecting containers for dwarf citrus, apples, and pears, consider the size and material closely. The container must be large enough to accommodate the tree's root system while having room to grow.

A 10 to 15-gallon (37.85-56.7 L) capacity is a good starting point, but some larger dwarf varieties might require up to a 20-gallon (75.7 L) container.

Ceramic or wooden containers are excellent for insulating the roots from extreme temperatures and are beneficial for heat-sensitive citrus trees, although these materials can be heavy. Always make sure your containers have enough drainage holes to prevent waterlogged soil.

Soil Mix and Fertilization

A well-draining, loamy soil is ideal for dwarf fruit trees to prevent water from pooling at the roots, which can cause root rot. Mix in a slow-release fertilizer formulated for fruit trees to provide steady nutrients throughout the growing season.

Additionally, ensure the soil pH is suitable for your tree type: Citrus prefers slightly acidic soil, while apples and pears thrive in neutral pH soil. Use a soil pH tester to ensure the soil remains within the ideal range for your specific fruit tree.

For mulching, you moderate soil temperature, suppress weeds, and retain moisture with mulch layered around the base of the tree.

Care and Maintenance

Dwarf fruit trees usually require minimal maintenance. However, it's still important to pay close attention to the needs and overall health of dwarf fruit trees to ensure they are not infested with pests, sick with a disease, or lacking necessary nutrients or water.

- **Watering Routine**: Properly watering dwarf fruit trees is important since their soil can dry out quickly, especially in warm weather.
    - **Newly Planted Trees:** Water deeply once or twice a week to ensure the roots are well-hydrated.
    - **Established Trees**: Should be watered deeply every 10-14 days, adjusting the frequency during hot spells to keep the soil moist but not waterlogged.
    - **Container-Grown Trees:** These may need more frequent watering, so check the soil regularly and water thoroughly when the top 2 inches (5.08 cm) are dry.
- **Lighting Management**: Position containers in a sunny spot where the trees can receive at least 6-8 hours of direct sunlight daily. This is especially important for fruit production.

    - Consider using containers with wheels or placing them on a movable platform to transfer trees easily for optimal sun exposure or to protect them from extreme weather.

- **Pruning:** Maintain your dwarf trees' shape and size to encourage healthier fruit production.
    - **Dwarf Apple and Pear Trees:** Prune them when they're dormant during the late winter to remove any dead or diseased wood and open the canopy to light and air.
    - **Dwarf Citrus Trees:** Can be lightly pruned year-round to shape, but avoid heavy pruning as it can reduce fruit production.
- **Fruit Thinning:** This technique complements pruning practices. Thin excess fruits early in the season to prevent branches from breaking under the weight and improve the remaining fruits' size and quality.
- **Pest Management:** Regularly inspect pests and diseases and take prompt action to keep your trees healthy and productive.

Pollination Needs

Dwarf apple and pear trees often require cross-pollination, meaning they need another variety nearby to set fruit. You can solve this issue by planting two different but compatible varieties. You can consider dual-grafted trees with two varieties on one rootstock for limited space.

Citrus trees are generally self-pollinating, but ensuring enough light and occasionally shaking the branches can improve pollen distribution and fruit set. Hand-pollination using a small paintbrush to transfer pollen may be necessary for indoor or sheltered trees.

To attract pollinators, plant flowers that attract pollinators like bees near your fruit trees to enhance natural pollination. You can check Chapter 5.4 to learn how to attract them safely.

On the other hand, to do artificial pollination for indoor or patio-grown trees, gently tap the branches or use a fan to simulate wind, helping distribute pollen.

7.3 Berry Bushes

Growing berries in containers offers a delightful opportunity to cultivate a range of berries in your urban space, whether on a balcony, patio, or indoors.

Success lies in selecting the right varieties and understanding their specific care requirements to ensure bountiful harvests. Compact and thornless varieties of blueberries, raspberries, and blackberries make the gardening experience fruitful and manageable.

Berry Variety Selection

Select berry bush varieties that are known for compact growth and manageable size.

- **Dwarf Blueberry Varieties:** 'Top Hat' or 'Jelly Bean' are ideal for containers.
- **Dwarf Thornless Blackberry Varieties:** 'Apache' or 'Arapaho' provide pain-free maintenance and harvesting.
- **Dwarf Raspberry Variety:** 'Ruby Beauty' thrives in pots with their contained growth and good yield.
- **Dwarf Wolfberry Varieties:** 'Goji Berries' or 'Crimson Star' are known for their compact size and high antioxidant content.

These selections are practical for space-limited gardeners and tend to mature faster.

Container and Soil Mix

Make sure to choose containers that are large enough for robust root systems. Blueberries need larger pots, about 18-24 inches (45.72-60.96 cm) in diameter, and thrive in acidic soil with a pH of 4.5 to 5.5. Use a mix specifically formulated for acid-loving plants.

Raspberries and blackberries will require well-draining soil, which can be achieved with a loamy potting mix containing organic compost, perlite, or vermiculite. These berries do well in containers at least 15-20 inches (38.1-50.8 cm) deep.

Make sure that containers have multiple drainage holes to prevent waterlogging and layer the mulch, retaining moisture and suppressing weeds.

Watering and Feeding

Maintaining the right moisture level is crucial. Blueberries need consistent moisture and watering when the top inch of soil is dry. Provide proper drainage to avoid root diseases. Raspberries and blackberries are a bit more tolerant but still need regular watering. Berry bushes are a bit more sensitive to changes in pH, so you can use rainwater or distilled water for berry bushes like blueberries to avoid high pH levels from tap water.

As for fertilizer, use an acid-formulating fertilizer for blueberries in early and late spring. Raspberries and blackberries benefit from a balanced fertilizer in early spring as fruits begin to form. Apply a slow-release fertilizer to maintain nutrient levels throughout the growing season.

Pruning and Winter Care

Each berry bush requires special attention due to its different pruning requirements and schedules. For blueberries, prune in late winter to remove old branches and encourage new growth. Raspberries and blackberries require specific pruning based on their fruiting habits.

Prune raspberries twice a year, removing spent canes after harvest and thinning summer canes in late winter. For blackberries, remove fruiting canes after harvest. Protect containers from extreme cold by wrapping them with insulating materials or moving them to sheltered locations.

You can successfully cultivate delicious berries in containers by understanding these specific needs and choosing the right varieties and care techniques.

7.4 Exotic Fruits

Dwarf fruit trees, vegetables, beautiful flowers, and herbs aren't the only things that can be grown in container gardens. You can also start growing exotic fruits like pineapples and pomegranates.

Exotic fruits add a touch of adventure to your gardening endeavors. Like most plants, these fruits require specific care and attention, but the rewards are, and will always be, worth the effort. For this next section, we'll focus primarily on pineapples and pomegranates.

Growing Pineapples

Cultivating exotic fruit can be rewarding, especially when you're growing sweet, tangy ones like pineapples. However, it does take a bit of patience when you're planting them.

Pineapples, commercially, are normally bred to be seedless, making their seeds a bit harder to come by. However, you don't really need the brown or black seeds to grow pineapples—you can simply make use of the pineapple's top/crown to grow more.

That, and it does take significantly longer to grow pineapple produce if you *do* plant them as seeds. So, for this section, we'll focus more on planting and growing pineapples by their crowns.

Starting and Planting

You can start by removing the top of a store-bought pineapple by twisting off the top, but if it's too difficult to twist, then you can cut off the leafy top with about an inch of fruit still attached to the crown.

If you do cut the top, trim away any excess fruit until you're left with the base of the pineapple leaves. Doing this lowers the chances of rotting.

After that, carefully remove several layers of the lower leaves to reveal approximately an inch of the stem. This exposed section is where the roots will develop.

Next, let the top dry in a well-ventilated area that has access to direct sunlight for about 2-7 days to prevent rot and heal cuts. To test the dryness, simply touch the cut and check if the exposed stem is no longer moist. Make sure that the leaves are still green and healthy-looking.

To plant the dried pineapple crown, prepare a potting mix, ideally a combination of sandy, loamy soil that's well-draining and slightly acidic with a pH of 4.5 to 6.5. Dig a deep, small hole at the center for the crown base and insert it.

Finally, ensure that the soil covers up to the base of the leaves and press it gently around the crown.

Care and Maintenance

Once the pineapple crown is planted, caring for it is quite simple. Here's what you need to consider in growing them:

- **Location:** Place the container in a warm, sunny spot, as pineapples thrive in heat and light.
- **Sunlight and Temperature**: Pineapples need at least six hours of direct sunlight every day. If you live in cooler areas, keep the plan indoors during colder months.
- **Watering**: Water regularly, especially in warmer months, but ensure the soil is not waterlogged—misting the leaves is especially beneficial since pineapples thrive in high humidity.
- **Feeding**: Feed pineapple plants with a balanced, water-soluble fertilizer every few months.

Do these steps consistently, and make sure to check for diseases and pests. Pineapple crowns, depending on the growing conditions, take about 2-3 years to grow.

Harvesting and Storing Pineapples

The plant will start flowering after 18-24 months and will take about six months to mature.

By then, check for these:

- **Check Ripeness: Pin**eapples have a specific characteristic that you need to pay attention to before harvesting:
    - **Color**: Look for color change—mature pineapples change from green to yellow-orange at the base, and the fruit should be at least two-thirds yellow.
    - **Scent**: It should start to smell sweet and fragrant at the base.
    - **Hardness**: Give the pineapple's side a few gentle taps. If it's hard, it's not ready. When it sounds hollow, then it's ripe and ready for harvesting.
- **Harvesting:** Prepare a sharp knife or a pair of garden shears to cut the fruit.
    - **Procedure**: Firmly grasp the pineapple at its base, close to where it connects to the stalk. Then, carefully cut the fruit off about an inch above the top of the fruit while making sure you don't damage the plant's crown.
- **Storage**: Once fully ripe and harvested, you can refrigerate the pineapple to extend its shelf life or enjoy it immediately.

When growing pineapples, there are some particular circumstances that need to be observed closely:

- **Temperature Sensitivity**: These plants thrive in temperatures above 65°F (18.3°C). However, they need protection or indoor placement during colder months.
- **Ripening**: If the pineapple needs further ripening, leave it at room temperature for a few days.
- **Propagation:** Pineapples can also be grown from slips or suckers, which are side shoots from the main plant.

Pineapple suckers can take about a year, or a year and a half, to grow.

- **Pests:** Watch out for mealybugs, scales, and other pests; treat infestations promptly with insecticidal soap or neem oil.

Growing Pomegranates

Pomegranates are elegant fruits that have small, ruby red, tear-shaped edible seeds inside called arils, which, unlike pineapples, you can harvest hard seeds from. The hard seeds are coated by a soft berry-like exterior, which is juicy and tarty when bitten on.

You can also collect these hard, crunchy seeds from commercially available pomegranates and grow them, but it will take at least six years for normal pomegranates to reach maturity. If you want to speed things up a bit, then growing pomegranates by cuttings or branches is a slightly faster option, taking at least three years to grow to maturity by cuttings instead of six.

For this section, we'll focus more on the dwarf pomegranate variety called "Nana," which is better suited for container growth. Dwarf pomegranates also take at least two to three years to reach maturity if you grow them via cuttings.

Growing pomegranates will also take a bit more care and patience, considering the amount of time it takes to foster and cultivate them. It can be a bit more complicated than growing pineapples.

Selecting and Starting

Before you begin gathering dwarf pomegranate branches from a nursery to grow in your own garden, it's best to make sure that you're taking cuttings during the spring when the dwarf pomegranate is actively growing.

Once it's spring, you'll need to prepare these materials and tools to make sure that the dwarf pomegranate grows healthily and properly:

- Pruning shears or scissors
- Rooting hormone powder (optional but recommended)
- Small pots (4-6 inches, or 10-15 cm, in diameter)
- Well-draining potting soil
- Plastic bag (for humidity)

After you finish gathering up the materials and tools, finding a healthy branch is the next step.

Be sure to choose 6-10-inch-long semi-hardwood branches from the current year's growth, meaning branches that have grown during the current growing season. They're light brown, or usually green, and are more flexible than the older branches of dwarf pomegranates.

Soft, flexible, semi-hardwood branches are ideal for cuttings. This is also the type of wood that is more likely to root successfully. Make sure it's free from disease and pests.

Use pruning shears or scissors to cut at a 45° angle and remove all leaves except the top two or three leaves. Removing the leaves reduces water loss, prevents mold, and encourages root growth, and leaving a few leaves at the top allows for the roots to continue photosynthesizing through those leaves; this way, the branch will have enough energy without losing too much water. The next few steps are more straightforward:

- **Apply Rooting Hormone**: Dip the part of the branches that you've cut into the rooting hormone powder and shake off any excess powder.
    - **NOTE**: This is optional, but it does help with root growth.
- **Prepare the Pot**: Fill small pots with well-draining potting soil and make a hole with a stick or your finger for each pot.
- **Planting**: Insert the cut branches into the holes of each small pot, and make sure that at least one node is below the soil level. Gather and press the soil firmly to ensure it's in place and doesn't fall over.
- **Water and Cover**: Water the cutting *immediately* after planting, keeping the soil moist but not soggy. Use the plastic bag to cover each pot to maintain humidity while allowing some ventilation to prevent mold.
- **Location**: Place these pots in a warm, bright area with indirect sunlight, as the cuttings will be too sensitive to direct sunlight.
- **Monitor and Care**: Keep the soil moist and mist the cuttings with water once or twice daily to maintain humidity. You should see new growth after 4-6 weeks, indicating the cuttings have rooted.
- **Transplant**: Once they successfully root and develop new growth, you can transplant the cuttings into larger pots of about 18-24 inches (45.72-60.96 cm) or, if you have a larger space, directly into your garden.
    - **NOTE**: Make sure it's a cool and cloudy day when you're transplanting the dwarf pomegranates to minimize shock.

It takes about 1-3 months for dwarf pomegranates to root successfully. Once they're rooted, it's time for consistent management.

Care and Maintenance

Once your pomegranate plants are established, follow these care tips to ensure healthy growth:

- **Location**: Place the plant in a sunny spot where your dwarf pomegranate can get at least 6 hours of direct sunlight daily.
- **Watering**: Water the plant deeply once every week to ensure that the soil dries out slightly between waterings, as they could develop root rot if they are overwatered. Make sure to water them on consistent days—if you water them on a Sunday, then continue watering them every Sunday.
    - **Hotter Climates**: You may need to water more frequently, once every 4-5 days instead of once a week.
    - **Colder or Humid Climates**: Once every 10-14 days is sufficient.
- **Checking Soil Moisture**: Stick your finger 2 inches into the soil. If it feels dry at that depth, it may be time to water. You can also opt for moisture meters as they're more precise.
- **Feeding**: Balanced, slow-release fertilizers are best added to your dwarf pomegranates in the early spring and again in the late summer.
- **Pruning**: Prune the plant annually to remove dead or crossed branches and to shape the plant. This encourages better air circulation and fruit production.

Like other plants, maintaining a consistent watering schedule and caring for your dwarf pomegranates can give them a higher chance of growing successfully. In terms of growth, it takes 1-2 years for them to grow foliage and develop a stronger root system.

Harvesting and Storing Dwarf Pomegranates

The flowering stage happens after 2-3 years as a sign of maturing. Six months after it flowers, which would have been about 3-4 years, the fruits will start developing and ripening.

Pomegranate fruits ripen for about three months, and the best time, usually, to harvest pomegranates would be from late summer to early fall.

Here are some characteristics that you need to watch out for in ripened fruit:

- **Check Ripeness:** Always check for ripe traits.
    - **Color:** dwarf pomegranate fruits should be a deep, vibrant red
    - **Scent:** These fruits have a subtle, sweet, and slightly tangy fragrance.
    - **Hardness:** Ripe dwarf pomegranates should be firm but not rock hard and feel smooth—the fruit should have a slight give when gently pressed. If it's too soft, then it's overripe. If it's very hard, then it's not ready.
    - **Sound:** If you tap it, ripe dwarf pomegranates could give a metal-like, hollow sound. If it sounds dull or flat, it might not be ready.
- **Harvesting:** Use a pair of sharp pruning shears.
    - Procedure: Cut the fruit from the tree, leaving a small stem attached to prevent damage.

- **Storage:** Store pomegranates in a cool, dry place. They can last for several weeks at room temperature or up to two months in the refrigerator.

Like any other fruit, pomegranates also have their fair share of traits that need to be looked for to grow successfully:

- **Temperature Sensitivity:** More tolerant of temperature variations but still prefer warmth and sun.
- **Propagation**: Pomegranates can also be propagated from seeds, but this method takes longer. For quicker results, use cuttings or purchase young plants.
- **Pests:** Aphids, whiteflies, and leaf spots can be frequent issues for dwarf pomegranates. Neem oil is useful against whiteflies and aphids, and having good airflow and avoiding overhead watering can easily prevent leaf spots.

Other Exotic Fruits

Pineapples and pomegranates aren't the only exotic fruits you can grow.

Here's a short list of other exotic tropical fruits that you can grow in your container garden:

- **Kiwis:** Miniature kiwis like *Jenny* or *Annas* are specifically bred for container gardening. Like most exotic fruits, however, these thrive under full sunlight.
- **Star Fruits:** Also known as *Carambola*, these fruits are visually appealing and taste sweet and tarty. A known miniature variant of this fruit is the "*Maher Dwarf*." These bright yellow star-shaped fruits will add to your garden's overall beauty.
- **Guava:** If you want an even healthier option, guavas are filled with vitamins and antioxidants. Variants you can grow are the common guava or the strawberry guava.

7.5 Pollination Tips

Understanding pollination can significantly boost the productivity of your fruit-bearing plants. Pollination—from a flower, an insect, or an animal—to a plant is essential for producing fruits and seeds.

Plant-to-Plant Pollination

Some plants are self-pollinating and can fertilize independently, while others rely on cross-pollination, which enhances genetic diversity and often leads to better fruit sets and quality.

- **Self-Pollinating Plants:** Plants like tomatoes and peppers have flowers with both male and female parts, allowing them to pollinate without external assistance.
    - Gently shaking the plant or tapping the flower cluster can help release pollen and improve fruit production.
- **Cross-Pollinating Plants:** Apple and pear trees, for example, require pollen from a different variety to set fruit effectively. This ensures a diverse gene pool, resulting in healthier plants and higher-quality fruits.
    - Planting multiple varieties of compatible species close together is key for optimal pollination.

Natural Pollination

Natural pollination is when your plants receive pollination from the existing local biodiversity within the area.

To encourage this for your container garden:

- **Create a Natural Pollinator Habitat:**
    - To attract bees, butterflies, and birds, grow pollinator-friendly flowers like lavender,

marigolds, zinnias, and sunflowers among your fruiting plants.

- Providing a consistent water source, such as a birdbath or a shallow dish with stones for insects to land on, will encourage pollinators to frequent your garden.

- **Cultivate and Seasonally-Rotate a Variety of Flowers:**
  - Have flowers that bloom at different times throughout the year to provide a continuous food source for natural pollinators.
  - Avoid using pesticides, as they can harm beneficial insects critical for pollination.
- **Enhance the Habitat:**
  - Introduce native plants to your garden, as they are well-suited to support local pollinators.
  - Incorporate features like nesting boxes for bees or logs for beetles to create a thriving ecosystem that benefits your plants.

Manual Pollination

Hand pollination is necessary in areas where natural pollinators are scarce, such as indoors or on balconies. Signs that plants need manual intervention include abundant flowers but no fruit set, which usually indicates poor pollination. By then, you can intervene and manually pollinate your plants to encourage fruiting.

To perform hand pollination:

- Use a small paintbrush, cotton swab, or even your fingertip to transfer pollen.
  - **For Self-Pollinating Plants:** Gently brush the inside of the flower to mimic a bee's movement and release pollen onto the stigma.

  - **For Cross-Pollinating Plants:** Collect pollen from one flower and carefully apply it to the stigma of another flower on a different plant.

Using this pollination method can be a delicate process, so it's best to be mindful of these best practices:

- Perform this task during mid-morning when flowers are fully open and pollen is most viable.
- Be gentle to avoid damaging delicate flower structures.
- Repeat the process every few days to ensure consistent pollination throughout the blooming period.

Indoor and Balcony Garden Pollination

When living in an apartment or a home with limited space, it's still possible for your container garden plants to pollinate as long as these factors are optimal:

- **Increasing Air Circulation:** In indoor environments where wind is limited or absent, use a small fan to simulate natural air movement. This helps to spread pollen between flowers effectively.
- **Supplementing Natural Pollination:** Maximize your plants' exposure to natural light and place them near open windows or doors to invite pollinators.
  - If possible, grow a mix of flowering plants and fruiting varieties to create an inviting space for insects.
- **Innovative Tools:** Consider using electric pollinators designed to vibrate flowers and mimic the buzz of bees. These tools are particularly effective for crops like tomatoes and peppers.

Pollination is essential to a successful garden, and understanding the process allows you to take proactive steps to ensure your plants thrive.

By combining natural pollinator attraction, manual techniques, and other innovative solutions through gentle application, you can create a container garden that flourishes with an abundance of fuller and healthier organic fruits and vegetables, all of which can give you a sense of fulfillment once it's time to harvest your produce.

Remember, you can always review Chapter 4.4 on Wildlife-Friendly Gardens to attract native, natural pollinators that can help your container garden thrive and allow you to reap more produce like fruits and vegetables.

Up next is the book's last chapter, which will cover more advanced techniques experienced container gardeners use to grow healthy, efficient, and abundant gardens.

Chapter 8

Advanced Techniques

You've gone through the previous chapters to understand and learn the different aspects of container gardening. And now, you're ready for the more advanced gardening techniques to grow, maintain, and care for your plants efficiently.

The first high-level skill you need to learn is grafting, a horticultural practice that allows you to blend the best traits of different plants into a single, more robust specimen.

Whether you're looking to improve disease resistance, boost fruit yields, or control the size of your garden plants, grafting opens a realm of possibilities that exceeds ordinary gardening practices.

8.1 Grafting

Grafting can be described as a form of living sculpture, a way to join parts from two or more plants by making clean cuts and binding plant parts so that they grow as one. The part of the graft that provides the root system is known as the rootstock, while the portion that grows into the trunk and branches is the scion.

Gardeners and farmers have been doing this for centuries because it helps plants fight diseases by combining the hardy plant features of one plant with the desirable traits of the other while having significantly more fruit yield. Grafted plants also ensure a more efficient delivery of nutrients. As a plus, if you have a small garden, grafting lets you control the size and shape of your plants so you can grow different fruits and vegetables in smaller spaces.

Grafting Materials and Tools

To start grafting, you'll need specific tools and materials to ensure success:

- **Grafting knife:** For clean cuts that heal well.
- **Grafting tape or strips:** Securely bind the graft junction.
- **Grafting wax or sealant:** Protects the grafted area from dehydration and infection.

Prepare both the rootstock and scion before you begin; make sure the scion is healthy and disease-free, with several buds along its length. The rootstock should also be healthy and well-watered.

Step-by-Step Grafting Processes

The grafting process might seem daunting, but with a step-by-step approach, even novice gardeners can achieve great results.

- **Cleft Grafting:** Start by making a vertical cut in the rootstock, creating a split about one to two inches deep. Shape the end of the scion into a wedge. Insert the scion into the cleft of the rootstock so that the cambium layers align. Secure the graft with grafting tape and cover it with grafting wax.

- **Whip Grafting:** Make diagonal cuts on the rootstock and scion, then fit them together so the cambium layers meet. This method is particularly effective for young, slender plants.
- **Bud Grafting (Chip Budding):** Transplant a single bud onto the rootstock. This method is widely used in fruit tree propagation and tends to yield successful results in late summer.

Aftercare for Grafted Plants

Don't forget that post-grafting maintenance is vital for the rootstock and scion to survive grafting:

- **Watering:** Keep the grafted plant well watered, but avoid overwatering.
- **Monitoring:** Regularly check the graft union for signs of growth or failure.
- **Support:** Provide support structures to stabilize the new growth.
- **Pest and Disease Control:** Watch for signs of disease or pest infestation and treat promptly.

8.2 Topiary and Espalier

The next skill you could learn is choosing the right plants for topiaries and espaliers.

Choosing Topiary and Espalier Appropriate Plants

For topiaries, small-leaved evergreens like boxwood or privet are popular due to their dense foliage and slow growth. Herbs like rosemary and lavender can also be topiarized, adding fragrance to your creations.

Meanwhile, fruit trees like apples, pears, and figs are ideal for espaliers. These trees lend themselves well to being trained against flat surfaces, which benefit from increased sun exposure and lead to more fruitful harvests.

Techniques and Form Conditioning

Topiaries and espaliers need different care and maintenance to keep their form and design stable and tidy.

- **Topiaries:**
    - **Framing:** Start with a form or frame that serves as a guide for shaping the plant. This frame can be as simple or complex as your desired topiary design.
        - **NOTE:** If you're still starting out, choose a sphere-shaped wire frame.
    - **Maintenance:** Your chosen topiary design will be regularly pruned and tied back to new growth to encourage it to fill out the frame. Begin with a young, flexible plant to easily train and shape it.

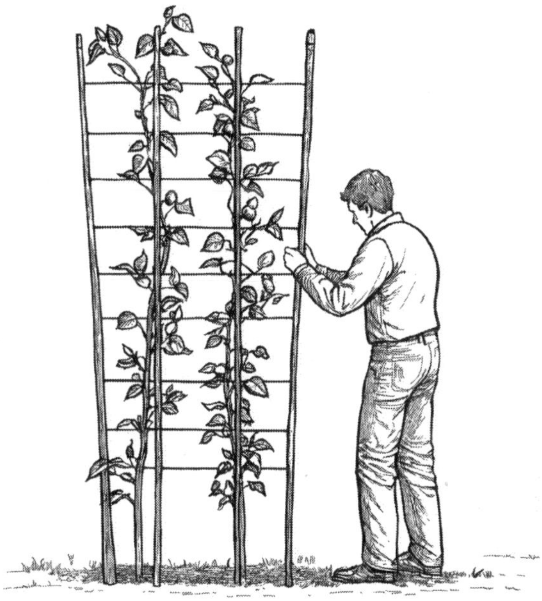

- **Espaliers:**
    - **Support and Trellises**: Espaliers would need sturdy supports like trellises or horizontal wires. These structures help guide the branches and ensure they grow flat against the surface.
    - **Pruning and Prep:** Next, prune and tie branches to the wires for maintenance, encouraging growth along the desired pattern.
        - **NOTE:** Regular adjustments ensure the plant maintains its shape and stays healthy.
- **For Topiaries and Espaliers:**
    - **Fertilization:** Careful fertilization management promotes steady, controllable growth without overwhelming the plant.
    - **Pest Control**: Regular inspections and timely treatments help maintain the health of your plants.

- **Pruning**: Regular pruning is essential for maintaining shape and health.
  - **For Topiaries**: pinch back new growth to refine the shape.
  - **For Espalier**: cut back fruiting spurs to enhance fruit size and quality.

Adding topiaries and espaliers into your container gardening allows you to make efficient use of limited space while expressing creativity through living plants.

Whether it's the intricate angles of a beautifully espaliered apple tree or the playful form of a topiary rabbit, these living sculptures provide a dynamic element to your garden that evolves with your gardening skills.

8.3 Using Greenhouses and Cold Frames

Embracing season extenders like greenhouses and cold frames can significantly strengthen your container gardening experience.

These structures can extend the growing season so you can start cultivating plants earlier in the spring and continue harvesting well into the cooler months.

This is especially beneficial for other gardeners who have to deal with limited space and challenging climates.

By creating a controlled microclimate, greenhouses and cold frames protect delicate seedlings, providing warmth and security from frost and chilly days and maintaining a more consistent temperature, which is crucial for germination and growth.

Building Season Extenders and Management

There are two structures that are commonly used as season extenders: cold frames and greenhouses.

Cold frames are simple structures that can be made using old windows or transparent polycarbonate sheets affixed to a wooden frame. Position them facing south or towards the sun to maximize sunlight exposure and facilitate rain runoff.

On the other hand, small greenhouses are more permanent structures that can be constructed with PVC pipes or wood and covered with clear plastic sheeting or polycarbonate panels.

You can personalize and configure the interior with hooks and shelves to make the most of your space and include adjustable vents or fans for air circulation.

To ensure that your season extenders are ready to cater to your plants' survival, regulating the temperature is a must.

- **Ventilation:** Automated or manual venting systems can help regulate the internal temperature. Use shade cloths to prevent overheating on sunny days.
- **Thermal Mass:** To maintain warmth, use water barrels or stones to absorb heat during the day and slowly release it at night.
- **Humidity Control:** Regularly open vents or doors to manage humidity and prevent mold and fungus.

Once your season extenders are built and prepared to foster plants, you can start planning out the seeds and plants you want to grow for each season.

Here are some general tips on how your season extender strategy might look from spring to fall—with winter being the time when you may continue growing your garden with hardier plants:

- **Spring:** Start seedlings in a protected environment before transplanting them outside.
- **Summer:** Shelter heat-loving vegetables like tomatoes and peppers.
- **Fall:** Extend the productivity of your garden by protecting herbs and greens from the cold.

Conclusion

Congratulations on reaching the conclusion of this container gardening journey! I'm truly excited for the steps you're about to take in creating your own thriving container garden. Whether you're a novice or an experienced gardener, I hope this book has provided valuable insights and inspiration for your gardening endeavors.

We've explored the essentials of container gardening, from selecting the right containers and preparing nutrient-rich soil mixes to choosing plants that suit your climate and space. Along the way, we've delved into maximizing small spaces and adopting sustainable practices that enhance your garden and support a healthier environment.

Container gardening is more than just a solution for limited space—it's an opportunity to connect with nature, cultivate creativity, and embrace sustainability. By incorporating eco-friendly techniques and experimenting with new ideas, you're not just growing plants—you're contributing to a greener planet and a mindful lifestyle.

Remember, gardening is a lifelong journey of learning and growth. Celebrate your successes, whether it's your first sprouting seed or a bountiful harvest, and approach challenges with patience and curiosity. Each experience refines your skills and deepens your appreciation for the natural world.

Thank you for allowing me to be a part of your gardening adventure. May your container garden bring you joy, satisfaction, and a deeper connection to the world around you.

Happy gardening, and best wishes for your continued success!

- *Kiley Manning*

Grow Together, Share the Harvest

Pass Your Knowledge Forward

"*The glory of gardening: hands in the dirt, head in the sun, heart with nature. To nurture a garden is to feed not just the body, but the soul.*" - Alfred Austin

Now that you've unlocked the secrets of container gardening, it's time to help others do the same.

By leaving a review, you'll guide other budding gardeners to the resources they need to cultivate their own lush gardens. Your honest feedback could encourage someone to begin their own journey and discover the joys of growing, just like you have.

Every review helps nurture this community of gardeners—let's keep the growth going!

Thank you for planting the seeds of knowledge and for being a part of this gardening adventure!

Kiley Manning

References

6 Ways to Control Common Container Garden Pests. (2023, August). Black Girls Gardening in Containers. https://www.blackgirlsgardeningincontainers.com/post/6-ways-to-control-common-container-garden-pests

Bender, S. (2024, April). 34 Shade plants for beautiful pots. Southern Living. https://www.southernliving.com/garden/shade/shade-container-garden

Boeckmann, C. (2024, November). Growing Up! How to grow vertically in your garden. Almanac. https://www.almanac.com/how-grow-vertically-your-garden

Bova, K. (2017, July). Small Space, High Yields – —How To Grow More in Less Space. Nourishing Pursuits. https://www.nourishingpursuits.com/garden/small-space-high-yields-how-to-grow-more-in-less-space/

Braidwood, J. (2022, October 3). 10 Best materials for indoor planter Drainage. Braid & Wood. https://www.braidandwood.com/blogs/news/10-best-materials-for-indoor-planter-drainage

Brolis, L. B. (2018, January). Community-Scale composting at urban gardens and farms. BioCycle. https://www.biocycle.net/community-scale-composting-urban-gardens-farms/

Chappell, M., Davis, B., Pennisi, B., & Sullivan, M. (2022, May). Landscape Basics: Color theory. UGA Cooperative Extension. https://extension.uga.edu/publications/detail.html?number=B1396&title=landscape-basics-color-theory

Christensen, E. (2022, May). Gardening: How to Hand-Pollinate Fruits and Vegetables. The Kitchn.

Elsner, L. (2023, September). What plant containers are best for container gardens? Epic Gardening. https://www.epicgardening.com/plant-containers/

Espiritu, K. (2023, May). 6 DIY natural pest control recipes for your garden. Epic Gardening. https://www.epicgardening.com/natural-pest-control-recipes/

Family Handyman. (2024, October). How to build your own Self-Watering Planter. Family Handyman. https://www.familyhandyman.com/project/build-your-own-self-watering-planter/

Growing Strawberries in Hanging Containers: The Ultimate Guide. (2022, August). Burpee. https://www.burpee.com/blog/growing-strawberries-in-hanging-containers-the-ultimate-guide.html

Henkler, D. (2023, March). How to make a faux topiary. In My Own Style. https://inmyownstyle.com/my-green-thumb.html

How to choose the right plants for your climate. (2023, August). The Cutting Edge Landscaping. https://thecuttingedgenj.com/2023/08/how-to-choose-the-right-plants-for-your-climate/

Hu, B., & Beartrack, M. (2022, April). Introduction to vegetable grafting. OSU Extension. https://extension.okstate.edu/fact-sheets/introduction-to-vegetable-grafting.html

Ingui, M. (2025, January). The 5 Best watering Cans of 2025, Tested and Reviewed. The Spruce. https://www.thespruce.com/best-watering-cans-4158875

Johnson, W., Costaris, C., & Nitzsche, P. (2023, October). Container Gardening with Vegetables. Rutgers New Jersey Agricultural Experiment Station. https://njaes.rutgers.edu/FS055/

Judd, A. (2024, November). Organic pest control that really works. Growing in the Garden.

https://growinginthegarden.com/organic-pest-control-that-really-works/

Keith, J. (2021, March). Five ways to maximize Small-Space Vegetable Gardens. Black Gold. https://blackgold.bz/five-ways-to-maximize-small-space-vegetable-gardens/

MacArthur, A. (2024, November). 6 Vegetable Garden Planner apps compared. Food Gardening Network. https://foodgardening.mequoda.com/daily/garden-design/vegetable-garden-planner-apps/

McCullough, N. (2022, March 31). Using Espaliers in the Garden. Garden Design. https://www.gardendesign.com/trees/espaliers.html

Miller, L. (2023, March). How to create organic soil for your healthiest garden ever. Better Homes & Gardens. https://www.bhg.com/gardening/yard/soil/creating-rich-organic-soil-in-your-garden-281474979752803/

Mr. Tiny Farmer. (2023, September). Choosing the right LED grow lights for your indoor plants [Video]. YouTube. https://www.youtube.com/watch?v=nnbNG1bHU5c

Noakes, H. (2021, May). 15 easy organic fertilizers to make at home. Sustainable Holly. https://sustainableholly.com/15-easy-organic-fertilizers-to-make-at-home/

Reich, L. (2022, October). Root pruning container plants. Fine Gardening. https://www.finegardening.com/project-guides/container-gardening/root-pruning-container-plants

Ruvo, K. (2016, November). Container specialty varieties. Renee's Garden. https://www.reneesgarden.com/blogs/gardening-resources/container-varieties

Ryan, B. (2023, July). Overwintering plants in containers. PennState Extension. https://extension.psu.edu/overwintering-plants-in-containers

Sanders, K. (2024, February). How to build a DIY cold frame. The Spruce. https://www.thespruce.com/diy-cold-frame-6363024

Stephens, N. (2024, March). Container Garden Companion Planting Guide. Permaculture Apartment. https://www.permacultureapartment.com/post/container-garden-companion-planting

Talerico, D. (2023, August). How to set up drip irrigation for pots, wine barrels, containers. Homestead and Chill. https://homesteadandchill.com/drip-irrigation-pots-containers/

Vandana, K. (2024, December). Maximizing your Space-Small indoor gardening ideas. UrbanMali Network. https://urbanmali.com/blogs/maximizing-your-space-small-indoor-gardening

York, P. S. (2024, March). How to grow and care for a dwarf fruit tree. Southern Living. https://www.southernliving.com/garden/trees/dwarf-fruit-trees

Made in the USA
Monee, IL
10 September 2026

62356694R00103